Spells That Cure

Author

Brian Daniel Starr

Spells That Cure

ISBN-13: 978-0-359-07929-2

LCCN: 2018910803

Starr, Brian Daniel is the owner of this work with all rights and responsibilities related to this work.

Copyright August 31 2018 by Brian Starr

Printed in the United States of America

DEDICATION

To

Lovely Daughter

Gabriella Maria Starr

Who has given me the

Patience an fortitude to

Complete this work

Also Dedicated to the cause for

Sainthood for

Blessed Ermengard

and

Venerable Servant of God

Father Michael McGivney

Founder of the Knights of Columbus

And the People who wore the Star during the Political Upheaval in World War II.

Contents

Section One Curing Spells
[1]Cure Itch spell

Daniel is father of Jacob who is father of Joseph who posts his angel and is father of Mannasseh who is father of Ebenezer who receives Joseph's angel, and was father of Moses.

[2] Health Spell

Jacob is father of Naptoli who is father of Asiel who is father of the Angel Raguel who is Father of the Archangel of healing Rafael and his sister Sarah.

Rafael is the healing angel, and is an archangel. Jacob is Israel or the father of twelve Israelites and Naptoli is one Israelites. As Israelites get older they find they become Semites when it is known what Jacob may have done.

[3] Health Spell utilizing a dying person.

[9] Rafael is father of Gabael who is father of Aduel who is father of Hananiel who marries Deborah of Epraim who is mother of Tobiel who is father of Tobit who marries Anna who is father of Tobias who marries Sarah the brother of Rafael.

This is similar to the last spell, except the line continues. Notice Sarah is six generations Tobias Grandmother.

[4] The Cure Lonelyness Spell or Shine Spell

Bathsheba, David, please allow your love from generations of incest to channel to me and shine like a diamond. Your parents loved each other since birth, and thru love making, and each generation makes more shine. If you have abundant love I know sometimes you wish to share. Thank you.

[5] Cure for Appendicitis

Astarte had Nimrod who had a son who had Jared the Father of Oriah (Orian) the father of Kip whose line goes to Omar whose line goes to Heth, the father of Uriah, the father of Absalom the father of Anna who has Rehoboam and makes Abijah. First Invoke the Clouds of the Sky called the Heth. Then remember Astarte the Goddess and her lover Nimrod.

[6] Death Spell of Broken Neck or Make an Angel

Aaron is father of Ithamar is father of Eli.

This is commonly done when Samuel is read by a person.   The subject is made to fall, and his neck bone breaks.   So the doctor puts the head in a halo, or brace so is neck does not move. This spell may cause death, and healing the bone takes about three months, so the person becomes an angel with a halo for three months.

[7] Cure Death Spell of Broken Neck

Aaron is father of Ithamar is father of Eli is father of a daughter (Crystal, or Christina, or Chrissy) who marries the high priest Uzzi and is also named Ely.

This spell is used because Eli is the Levite priest that makes the fall, however has a daughter that is not really well known and so marries the Levite Uzzi and gets the person who receives the Death Spell into a different Priest.

[8] Name the Stars.

(Generals)

Zacheus First Star, or Adjajjent (one us)

Alphaeus, who make Two Star, or Brigidier. (two us)

Thadeus, the three star or Major General (Three Us)

Jonus. The four star or General or (four us)

Not an us. And some Idiot put on five stars, which doesn't make any sense, so he can go to hell. So the Hellenism of Saint Helen makes cures out of the hell left by the commanding General.

[9] Idiot Spells

There are not parallel realities, there are people with sophisticated computer programs that brain wash people with computers remote operation, and televisions, using subliminal suggestion and other intelligence tricks.   These people are dangerous idiots. Acknowleding them and paying any sort of attention is also very dangerous.

Idi Amien ruler of Uganda, to discipline his prince son took his wife and cut off all limbs and switched left to right and right to left.

[10] Limp Dick Spell

Astarte Ningal Enkidu, NinMah, Anu, Anskar, Tiamet, Apophis, wife of Yam, Iluyanka

This spell is the difference about a Yank.   Possibly the origen of the Yank or Yankee term.   Made from the Sumatrian dieties. This spell will cause the victim to be completely hasseled and bothered with while a lady massages, sucks, or anything and the victim is continually distracted an cannot concentrate enough to get it up.

[11] Counter Limp Dick Spell

Astarte Ningal Enkidu, NinMah, Anu, Anskar, Tiamet, Apophis, wife of Yam, Iluyanka daughter Babys, Perecide son of daughter of Ion, Ion, Sol or Apollo.

To counter the Limp Dick Spell you must get away from Iluyanka. And go to the Greek God Apollo son of Zeus. The Greater the Knowledge you possess of Iluyanka, her ascent, descent and other relationships the greater your counter will be.

[12] Spell of Division

When a deified God uses a hatchet to divide the black snake during the pregnancy of the mother subjected to more hel than any other women, who can assimilate the spirit of the serpent as well as the spirit of her noble father into the deified Gods Grandson, who is destined to make kingdoms topple and cure great diseases, stealing death from the powers that be and be more powerful than the possession the serpent uses, by possessing all gold and overcoming the pain of dispossession.

[13] Spells of Nursing

Rebecca Towne Nurse was an innocent lady, who was hung as a witch by Governor Phipps, in Salem Massachuesetts.   She was not a witch, she was not guilty of Killing Anyone.

Astarte, Ningal, Enkidu, Ninmah, Anu God of the Sky, Enlil God of the Wind, Ninlil the Beautiful Nurse

Rebecca had Twenty Four Cousins, including her cousin (half sister) Deborah her nurse daughter of Huz. In the Universal Faith Rebecca and Deborah have Three Letter Codes.   OEX is likely Deborah, written here of course in the English typing of the Aramatic language. To bring up past the Aramatic an asking of the passage to the true lettering might be heard if the guard deems it worthwhile.

[14] Spells to Cure Possessions for Demons for Mary Magdalene.

Now Mary Magdalene was possessed by seven demons according to the scriptures. These demons may be presented in the Gospel of Mary Magdalene that was found in the late 19th Century. These demons were darkness, Craving, ignorance, Lethal Jealousy, enslavement to the body, intoxicated wisdom, and guileful wisdom. There is a genealogy that explains the heresy and the truths and parallels that the heretical take advantage of.

The darkness curse is done with the iris of the eye. It can be judged to open or close to regulate the amount of light the subject sees. If closed enough the darkness can cause depression in the person due to the dim light.

Craving is difficult to counter, if about food possibly becoming a vegetarian, or if about sex, abstenance or satisfy the craving constantly until marriage or the lust is consumed.

The ignorance curse is the easiest to counter, however education or instruction takes a while.

Lethal Jealousy is difficult to contain, and if used will likely get someone killed. Not a good thing. Better to realize that if you really love someone that much then it is better to set them free because you love them.

Enslavement to the body is difficult, if its sex, ok, if its exercise, or if it's the idea you have to look good for your mate, then reason will likely repel this demon.

Intoxicated Wisdom is usually with alcohol or excessive work. To repel this demon either sober up or rest.

Guileful wisdom demon is repeled by humility, if you know you are not that smart then quit making arrangements for people.

[15] Spell to Resurrect the Dead

Dear Lord please resurrect the following person who has died (state the name) Post Christianity King Arthur who is blood of Heber Hyperion who marries Tamar the granddaughter of Josiah King of Judah who is a descendent of Jehosaphat a descendent of King David. Jehosaphat is married to Malika who is of the line of Gershom son of Levi. So from Malika navigate to Asaph who wrote many songs (straight up the line) who has many sons and is a boys choir, so the boys are the innocent and Michael the levite or Gershomite takes the dead soul from its carry and makes it part of the Pregnant Woman, whose child now has a soul that is both new and resurrected.

[16] Spell of Love Arimathea

The spell of Arimathea requires a Post of Saint Joseph of Arimathea and a correct veneration from Saint Caradoc of the Lord. These lines are hidden so that everyone who calls on the name of the Lord does not start a Arimathia.   Also Medical doctors may prescribe heart medicine for an irregular heartbeat. Before Casting this it is a good Ideat to Post Saint Anne, and also make sure the lover is healthy enough. Also not for Children . Most of this is for the Mother and Father of whoever so that the Arimathea will generate enough love to last thru the night.

Navigate to Saint Anne of Arimathea.   Check the Euygeus and go to Rachel of Arimathea. If She wants you to feel love of Arimathea, she will post Anne the Priestess daughter of Joidiah the Levite and then post the true Eugeus from Caradoc and the line of Jesus the crucified son of Alphaeus son of Bar Jesus Son of Saint Joseph of Arimathea.

[17] Spell of Blessing

Judah, Bathshua, Hasadiah, Zedekiah, Masseiah, Neriah, Baruch (Baruch in this language at this time means Blessed so his name is Blessed)

[18] Spell to Raise the Sun

Astarte, Ningal, Enkidu, Ninmah, Anu, Anskar, Tiamet, Apophis, daughter of Yam, Iluyanka, daughter, Baby, Perecides, son of Ion, Ion, Sol, or Apollo

[19] Spell to Raise Death

Jacobs son Levi, Kohath, Amram, Jocabed and Elizabeth mother of Nadab. (Death) with his brother Abihu.

[20] Spell to Raise the Staff

Ket, Serug, Nahor, Terah, Abraham, Isaac, Israel, Joseph, Manassas, Jepunteth marries Hezeon Jathir (Wood)

Ket, Serug, Nahor, Terah, Abraham, Isaac, Israel, Judah, Perez, Hezron, Jathir (Wood)

[21] Spell In the Knight to Add Food and Shelter to the Pregnant Women with Child

Mathan, Jacob, Eucheria, Miriam, Theudas, Addai James, Unknown Father of Sotor,

Unknown Selius goes to Ptolas or Bartelmaeus, and buys the beasts in Worship of the Man who proved he was god of that beast, and then to Ham and Olivia to Domnu to Add the Songs of Life to the child and then to Paratama the Overlord, and to Shaushatar the wood Lord and to Aratama the Horse Lord, and to Shuttarna the War Lord and to Tushratta the Axe Lord.

[22] Spell to Cure Arthritus

Saint, Saint, Saint, Joe of Arthimea in the right to Saint Joseph of Ariamathea

[23] Spell to Cure Cancer

Oh Great King and Saint Ethelbert Please cure this Cancer.

Saint Ethelbert believe in God Odin and ask the favor of God's Wife Frigg whose lineage marries Buonduah whose lineage marries Ann the Daughter of Jose the Rama Thea the son of the Saint Joseph of the Davidic Kingdom.   Saint Joseph post your daughter Ann and ask your grandfather Mathan favor with his daughter Don who marries Mathan Great Grandson Beli.

[24] Raise the Priest Spell in the Knight

Saint Joseph, let your father revenge your enemy son of Jacob who marries Eucheria mother of Miriam who marries Theudas, son of Anthrongonies the Good Shepard father of Amran

[25] Spell to Remove Negative Thoughts

O Great Father Abraham, remember thy name of God r n b

On the passage of the interpretation of the aromatic translation of the three letter Universal, please Lord bring the true interpretation to this Praying if found worthy.

[26] Spell to Arouse Total Certainty

O Great Father Abraham remember thy name of God rvk

On the passage of the interpretation of the aromatic translation of the three letter Universal, please Lord bring the true interpretation to this Praying if found worthy.

[27] Spell to Arouse Healing Powers

O Great Father Abraham remember thy name of God o e x

On the passage of the interpretation of the aromatic translation of the three letter Universal, please Lord bring the true interpretation to this Praying if found worthy.

[28] Spell to remove Negative Forces

O Great Father Abraham remember thy name of God L E Z

On the passage of the interpretation of the aromatic translation of the three letter Universal, please Lord bring the true interpretation to this Praying if found worthy.

[29] Spell to Generate the Energy of Financial Sustenance

O Great Father Abraham remember thy name of God Q A N

On the passage of the interpretation of the aromatic translation of the three letter Universal, please Lord bring the true interpretation to this Praying if found worthy.

[30] Spell to Remove Egomania

O Great Father Abraham remember thy name of God A P K

On the passage of the interpretation of the aromatic translation of the three letter Universal, please Lord bring the true interpretation to this Praying if found worthy.

[31] Spell to Eradicate Death

O Great Father Abraham remember thy name of God P K Z

On the passage of the interpretation of the aromatic translation of the three letter Universal, please Lord bring the true interpretation to this Praying if found worthy.

[32] Spell to Return to Seed Level of Existence

O Great Father Abraham remember thy name of God F E F

On the passage of the interpretation of the aromatic translation of the three letter Universal, please Lord bring the true interpretation to this Praying if found worthy.

[33] Spell to Stand After we Fall

O Great Father Abraham remember thy name of God E U B

On the passage of the interpretation of the aromatic translation of the three letter Universal, please Lord bring the true interpretation to this Praying if found worthy.

[34] 𝔖𝔭𝔢𝔩𝔩 𝔱𝔬 𝔍𝔤𝔫𝔦𝔱𝔢 𝔞 𝔖𝔱𝔞𝔯

Simple Algebra.

Note the equation to calculate the force of repulsion between a Proton and a Neutron.

$F = k*Q1*Q2/r^2$

Force is equal to a Constant k times the charge on particle 1 and the charge on particle 2 divided by the distance between squared.

$F = G*M1*M2/r^2$

F is force between masses.

G is the Gravitational Constant

M1 is the First Mass

M2 is the Second Mass

R is the distance between the centers of the Mass

Now the Trick   F=F

Force from the first equation is the same as Force from the Second Equation

Therefore the force of gravity between the particles will overcome the electron force to repel the proton.

$K*Q1*Q2/r^2 = G*M1*M2/r^2$

or

$M1*M2 = k*Q1*Q2/G$

Solve for M1*M2 or Total Mass   This is Roughly the Mass of Jupiter the Morning Star

[35] The Spell to Arouse a Man

Great Goddess, let your husband Amenhemet father of Sensuret II who marries Nefru father of Meribah let the water flow, while her mother marries Tjenna mother of Tetisheri who marries Semkearee Tao I the wolf. (the song of the wolf is very much like the song of man).

[36] Spells to Cure Lonelyness

O Great High Priest Aaron guide us to your Law Giving Brother Moses whose Father Amram' Brother Great Descendent Assir has a Great Descendent Assir whose son Kore thru Obed Edom is father of Sacar the Father of the Greatest Love Ahiam.

[37] Spell to Cure Erectile Disfunction to Make Erection

(125) Xerxes and Esther Jair know your Mordecau Jair know Ira

(126) Judah father your Son who fathers Mahari and your Son fathers Baanah who fathers Helad, and then your Son fathers Ikkesh Father of Ira

[38] Spell to make wife for a Lord

O Good King David let your might strengthen Zelek.   Let Abala know Nahash and father Shobi Father of Zelek Father of a wife for a Lord

[39] Spell to Calm a Women

ArchAngel Michael Find your arc from Gad, and let Mary know Hyper duluth and let the women rise to Heber Hyperion whose wife Grandfather is Good King Josiah. Let the Good King Josiah know his wife Zebudah and let her take the women to her father Pediah son of Jehiochin son of Queen Nahushta. Let the Good Queen take the women up her line to Elkanah thru his wife to the shelter of Gilead son of Michael.

[40] Spell To Enter the Heaven Olympus

Our Lady of the Cross lead us to the Worthy Warrior Emperor Alexander the Great and let his mother Olympias line rise to Neoptolemus son of Achilles and Iphigenia daugher of Helen Daughter of Zeus.

[41] Spell To Cure Gout

Our Father Abraham, know the son of God Isaac, father of Esau, Father of Dan, Father of Shobal, Father of Ammisidab, Father of Moses, Father of Gerson and his Half-Brother Awana, father of Piye, Father of Kashta, Father of Piye II who married Aqualqa, parents of Tirhakah, father of Ishtemebat who married Kenesat, parents of Psamtek and Mehetenweakhey, Parents of Necho, Father of Scota Tephi Nectaebus, father of Nebuchadrezzar and his brother Heremon Eochaidh Supra

[42] Spell To Cure Aids

Daniels Vision of the Beast

Lion Eagles Wings

Mystical

Beasts (see Book of Daniel for Details) Obviously a Tiger.

Leopard

With

Eight

Eyes

Mystical

Beast

With

Feet and

Kingdoms

Kingdoms

Are the

Four

Fathers

These are

Moab

Lot

Ablimelech

Abraham

The Lines to Sarah

Japeth

Gomer

Riphath

Bithiah

or

The Lines to Sarah

(The Lines divisions show the four mystical Beasts.)

First Beast.

Cham

Cush

Nimrod

Boethus

to

Second Beast

Raneb

Nynetjer

Peribsen

Nimaethap

Djoser

Third Beast

Sekhemkhe

Khaba

Huni

Sneferu

Knufu

Djedefhor

Fourth Beast

Khentkawes

Nefirirkare

Shepseskare

Neuserre

Menkauhor

Djedkare Wenis

Wenis

Iput

Pepy

Pepy II

Nitrocris

Ankhfn-Khonsu

Menuhotep I

Inyotef

Inyotef II

Inyotef III

Menuhotep I

Menuhotep II

Menuhotep III

Nefir

Sarah

Doctor

Nose

Common

Cold

Pnemonia

Throat

Mononeucleos

Strep Throat

Ear
Ear Ache my Eye

Eyes

Color Vision
Black and
White Vision

Adding How the Daughter Thinks.
She Reasons every one only has one mother. She really gets a headache if you argue about that. She's pretty smart.

First Beast

From Noah
Oliviana
Goyong
Bithia
Nemathap.

Second Beast

From Mother Nimathap

Hotepwirebti

Knufru

Nebet

Third Beast

From Mother Nebet

Nebet

Nebet II

Ankhenemerire

Merene Nemtpesf

Fourth Beast

From Merene Nemtpesf

IPU

Knenut

Ipwet

Ipwet II

NItrocris

Nitrocris II

Ankhfr

Abenra

Nuferukayet

Aoh

Nefuru

Imi

Nefir

Sarah

The First Second and Third Beasts have daughters that are from other Dynasties. These daughters marry again and then the line from Oliviana that leads to Sarah Joins at the end of each Beast.

So the line shown on the other page is a continuous line of daughter to mother starting at Sarah, that marries from Abraham to Noah.

There are other dynasties that go as well.

The Fourth Beast has a parallel that goes to Aseneth and the dynasties of Joseph. This is for Mary and Marriage and will heighten the lady if she choses to use this.

From Merene Nemtpesf

Ipu

Sitiah

Zelekha

Asenath.

So if the First Beast is the double p Pnemonea, as usual it would make little difference if the immune system was weakened by the common cold until the end, or if you just get double p Pnemonea and then its over.

Either way I suppose. So if the daughter knew the mother then she would know how to pass thru all this.   Unfortunetely the way is difficult for the men, as the marriage was to Irad, who did not mind taking an unholy vow and so remained on the dark side.
So please stay away from the Hittites if you are in the First Beast.

Please keep this in mind if you are in a position to make a serum or something that could cure this.

[43] Spell To Help Women Abused by Husbands or Boyfriends.

Also Starr Books has released the Venerations of Eliza Allen Starr. If anyone in the Vatican or Elsewhere would like to follow up on her Sainted Lists. Her books are reprinted (some are) and her original Books are rare collectors items. Her book Patron Saints is available and other books are available in rare book stores. Her List of Saints is over 100 Years old.

Many Venerations were made by Eliza Allen Starr in her lifetime, that ended about 1905 or 1906. She as a servant of God is used to help people like women that are either in danger of being abused or have been abused. Please ladies always keep a safe house or a place to flee to if the husband or boyfriend decides to become violent or in anyway abuse the lady.

The book as written are venerations of The Starr line that the author has in common with Eliza Allen Starr, our common ancestry joining ten generations up, each thru only the surname Starr.

If you feel so inclined to make these venerations a part of the Servant of God Eliza Allen Starr, then I see no reason why not to. It couldn't hurt, and in fact if might help someone.

The book is available and the venerations in some cases already work, as the author knows certain of these were set up for those of us with similar blood lines to Servant of God Eliza Allen Starr over a hundred years ago.

Section Two Spells Of Magick

[1] Sleep

Nahor had Iosaka who made Terah who had Maria who made Haran who had Yawnu who made Milcah who had the same Nahor and made Bethuel, whose half-brother Terah had had Maria who had made Haran who had Yawnu who made Micah who had Bethuel who made Rebecca.

The Precident is that Terah has to be older than his half brother Bethuel, because Terah is Bethuel's Great Grandfather.

Incidently Rebecca has 24 cousins, including her dad Bethuel.

This is called Bethany in most ways.   It is genealogy based, with a navigation from Nahor to Rebecca.   It helps if the caster has rested his head on a rock, endured the pain, and gone past feeling the pain.

[2] Itch spell

Daniel is father of Jacob who is father of Joseph who is father of Mannasseh who is father of Ebenezer who is father of Moses who is father of the prophet Malachi......

Incidently Joseph has an angel.

This is a part of a line that goes from Daniel Son of Abigail, to Saint Gamaliel.   It is watched due to the mention of Moses.

[3]Cure Itch spell

Daniel is father of Jacob who is father of Joseph who posts his angel and is father of Mannasseh who is father of Ebenezer who receives Joseph's angel, and was father of Moses.

[4] Death Spell

Mene, Mene, Tekel, Perez. (Mene, Mene, Shekel Phares.)

If you can't fart, and you can't burp, then in three days the pig inside you causes internal bleeding.

A Ghastly way to die.

This spell is found in the Hebrew book of Daniel.   It is cast by god and Daniel's keeper Beltzaahar dies.

The mene parts are full menestrations by women, the price is the Shekel, a type of coin the Hebrew use and the Phares is like the Pharoah or the priest.

[5] Dumb Spell

Causes dumbness or the inability to speak in the presence of a beautiful women

[6] Astarte had Zabada who made Shala who married Enlil (God) and then Shala had Iskur who made Dumuzi who had Astarte and made a daughter who had Gilgamesh.

This is a navigation from Astarte, who married Cush, his son Nimrod, and his son Gilgamesh, and Zabada and Dumuzi.

[6] Luck Spell

King Arthur line continued to Heber Hyperion who married Tamar the granddaughter of King Josiah, whose wife was the daughter of Pediah who was the son of Jecoliah who was the son of Jehoikim who married Nahusta of the line of Gad that married the line of Sharmariah Son of David.

This is the Shamrock spell, or veneration of the Son of King David.   Shamariah has descendents that marry the tribe of Gad. However at King Josiah, a Hebrew king, the ascent goes to King Joasiah, and then to his other wife, so this is a navigation not a veneration.

Strength Luck,   Mathan, Hizkiah, Judas the Zealot, Simon, Eleazar, Andrew Lukuas

[7] Health Spell

Jacob is father of Naptoli who is father of Asiel who is father of the Angel Raguel who is Father of the Archangel of healing Rafael and his sister Sarah.

Rafael is the healing angel, and is an archangel.   Jacob is Israel or the father of twelve Israelites and Naptoli is one Israelite.

[8] Health Spell utilizing a dying person.

[9] Rafael is father of Gabael who is father of Aduel who is father of Hananiel who marries Deborah of Epraim who is mother of Tobiel who is father of Tobit who marries Anna who is father of Tobias who marries Sarah the brother of Rafael.

This is similar to the last spell, except the line continues. Notice Sarah is six generations Tobias Grandmother.

[9] The Wheel spell (involves the Astarte Goddess Tara.)

TAROTAROTAROTARA

The Wheel in the above spell will post three O and then the most intelligent mind of Astarte Goddesses will interpret. Can be used for all commissioned rank, depending on how many ties the wheel goes around.

The Tarot has always been associated with Tara, who is usually gifted as a medium or astrologist if she decides to use this talent.

[10] The Earthquake Spell.

(spoken by Phillip at his crucifixtion)

And he began to curse them, invoking, and crying out in Hebrew: Abalo, aremun, iduthael, tharseleon, nachoth, aidunaph, teletoloi: that is, O Father of Christ, the only and Almighty God; O God, whom all ages dread, powerful and impartial Judge, whose name is in Thy dynasty Sabaoth, blessed art Thou for everlasting: before Thee tremble dominions and powers of the celestials, and the fire-breathing threats of the cherubic living ones; the King, holy in majesty, whose name came upon the wild beasts of the desert, and they were tamed, and praised Thee with a rational voice; who lookest upon us, and readily grantest our requests; who knewest us before we were fashioned; the Overseer of all: now, I pray, let the great Hades open its mouth; let the great abyss swallow up these the ungodly, who have not been willing to receive the word of truth in this city. So let it be, Sabaoth. And, behold, suddenly the abyss was opened, and the whole of the place in which the proconsul was sitting was swallowed up, and the whole of the temple, and the viper which they worshipped, and great crowds, and the priests of the viper, about seven thousand men,

The Church Fathers. The Complete Ante-Nicene & Nicene and Post-Nicene Church Fathers Collection: 3 Series, 37 Volumes, 65 Authors, 1,000 Books, 18,000 Chapters, 16 Million Words (Kindle Locations 162425-162429). Catholic Way Publishing. Kindle Edition.

The Church Fathers. The Complete Ante-Nicene & Nicene and Post-Nicene Church Fathers Collection: 3 Series, 37 Volumes, 65 Authors, 1,000 Books, 18,000 Chapters, 16 Million Words (Kindle Locations 162421-162425). Catholic Way Publishing. Kindle Edition. Ephratha ? ?

This is a dangerous spell and is cast when Phillip is undergoing public execution. The moral of the story here is do not go to a public execution.

[11] The spell of Ishmael

(made in the seventh month of a pregnancy, so she will make milk for her child) [Part of Gabriel]

Gabel to Penuil   break Ishmael.

There is a line in every man and woman that connects the tits, nobs, breast, gabes, etc.. to the hip. This connection is patrolled and if making Gabriel it is Gabriel. Ishmael is mothers milk but it takes a few months for the glands to transition to be suppliers of felicity to the baby.

[12] The Love Spell or Shine Spell

Bathsheba, David, please allow your love from generations of incest to channel to me and shine like a diamond.   Your parents loved each other since birth, and thru love making, and each generation makes more shine.   If you have abundant love I know sometimes you wish to share. Thank you.

[13] Spell to Sex Child to Preference

Lud Law to Caesars to the Caesar Flavius Sextus who is a descendent of Flavius Julius Augustus Caesar.

This spell is a veneration, keep in mind that Flavius Sextus is not the chosen line of the Emperors, but is a second line.

[14] Spell to Make living thing (can neither be destroyed or Altered) (Dominic)

[15] Songs of Life Spells

Olive had Ham whose child was the parent of the Wife of Domnu the son of Chaos and Nxthy the children of Caligo and Athys.

The song of Domnu is the song of life for all things God had made.

The Olive tree starts with car------

The lamb Domnu starts with Michaiah……..

The Corn Plant Domnu starts with Whitigen………

The Snake Donmu starts with Cleoputre…………

The Hela Monster Donmu starts with Gihon……………

[16] The spell of increasing the Psychyey or Psycopath or keep the children on the right Path.

Caligo had Athys who had Chaos Who had Nytx who had Erebos who had Nytx make Eros who had Psyche who made Voluptua.   (Voluptus Daughter)

Great God Zeus let your titan father Cronus remember his father Uranus whose mother was Gaia who married Chaos Who had Nytx who made Erebos who had Nytx make Eros who had Psyche who made Voluptua.   (Voluptus Daughter)

[17] The metallurgic spell of making Brass

Copper and Bronze make Brass at the forge.

[18] Control the speed of Time

Ithreel is the time stream   King Saul is father of Eglah who had Phatiel the son of Laish the son of King Saul.   Eglah also had King David who was the father of Ithream.

This spell is part of King David. It can be dressed up with Jupiter and Saturnus, for forward and backward travel in the Ithream time stream monitors by the King.

[19] The spell of fire (male and female)

Caligo had Athys who made Chaos who had Nytx who made Domnu whose line continued to DelBreath whose line continued to Donan who was mother of Brian, Iachar, and Iacharba.

Iachar, and Iacharba are fire man and fire women. Char makes fire.   Ia is like I am. Navigate to Chaos and cast away !!!

[20] Become a Wizard and Raise the Angel of Death

Ruth had Salomon and made Ibzan who was the father of Edal who was the Father of Abala who had Jesse but also Uriel the dark Angel   (Restores youth thru labor)

This is not known since Ruth is married to Boaz and Salomon is Boaz father.   So Ibzan is the wizard, and the line continues to Abala who married Jesse of Bethlehem, but slept with the levite Uriel who is the angel of death or the dark angel.

[21] Cure for Appendicitis

Astarte had Nimrod who had a son who had Jared the Father of Oriah the father of Kip whose line goes to Omar whose line goes to Heth, the father of Uriah, the father of Absalom the father of Anna who has Rehoboam and makes Abijah

[22] Death Spell of Broken Neck or Make an Angel

Aaron is father of Ithamar is father of Eli.

This is commonly done when Samuel is read by a person.   The subject is made to fall, and his neck bone breaks.   So the doctor puts the head in a halo, or brace so is neck does not move. This spell may cause death, and healing the bone takes about three months, so the person becomes an angel with a halo for three months.

[23] Counter Death Spell of Broken Neck

Aaron is father of Ithamar is father of Eli is father of a daughter (Crystal, or Christina, or Chrissy) who marries the high priest Uzzi and is also named Ely.

This spell is used because Eli is the Levite priest that makes the fall, however has a daughter that is not really well known and so marries the Levite Uzzi and gets the person who receives the Death Spell into a different Priest.

[24] Save from Death from Israel or Keep the death in Israel

The Jew-is

The 6th King of Judah Jehoram married the 7th King of Israel Ahab's daughter Athalia the 8th queen regent of Judah whose Brother Jehoram was the 9th King of Israel

The history is the Israeli King Jehu comes after Jehoram.   So the priest makes an attack on the subjects Judan Jehoram, however is countered by this spell and the Jehu harvest is done to Jehoram the King of Israel not Jehoram the King of Judah.   Then the covenant between Jesse of Bethlehem and Jeroboam is invoked that is that Jeroboam, and his descendents, will shepard and not kill Solomon's descendents, who are Kings of Judah.

[25] The Take Over the World Spell

The Tear or the Generals   Five Stars

Zacheus or Zachariah married Elizabeth and made John the Baptist.   Zacheus cheated and had Bernice, who was married to Herod Antipas, and so Herodias was born.   So John The Baptist liked Herodias and had her so his justice was done in the cell for incest

So Zacheus was sacrificed in the doorway by Alphy to revenge his wife's rape.

So Alphaeus married Herodias, who had Mathew, so the Virgin Alphy's half sister slaked Alphy and she made the lord Jesus, so Ann slew the Virgin for slaking and that was her Justice for Adulterous Incest. So Alpheus slew his mother for killing his lover.

So Alphaeus had Cleopus wife Mary because she slaked him and Mary made James the lesser and Judas Thadeus. So Herod Antipas slew Alphaeus for cheating on his daughter.

So Judas Thadeus revenged his father and slew Herod Antipas so the blood of King Herod was split so Herod Phillip slew Judas Thadeus to revenge his brother, so his brother James the Lesser slew Herod Phillip, so Herod Agrippa II slew James the Lesser to revenge his Uncle, so Herod Agrippa II had Bernice who made Perpetua.

Herod Agrppa II was married to Bernice But Simon had Bernice to revenge his brother James the Lesser who made Mariam Arrias who married Marcus Titus Flavius Sabinus so he slew Simon to revenge the rape of his wife's mother so Andrew slew Marcus Titus Flavius Sabinus to revenge Simon so Gaius Sillius Calpurnius Domitius Piso slew Andrew to revenge his father so Jonas slew Gaius Sillius Caopurnius Domitius Piso to revenge his eldest son.

So Jonus had Mariame Caecina Arria Sabinus who made John Mark, so Arrius Antonius Calpernius Piso slew Jonas for the wife being raped So Peter Slew Arrius Antonius Calpernius Piso to revenge his father Jonas so Bonionia Prossilla Servila slew Peter to revenge her husband.

So Bartholomew slew Bonionia Prossilla Servilia the wife of Arrius Antonius Capernius Piso to revenge Peter so Arrius Antonius Calpernius Piso slew Bartholomew to revenge Bonionia Prossilla Servilia his wife so James the Greater slew Arrius Antonius Calpernius Piso to revenge Bartholomew, so Marcus Annius Verus slew James the Greater to revenge Arrius Antonius Calpernius Piso his father so Phillip did Battle and slew Marcus Annius Verus to revenge James the Greater so his son Lucius Arrias Verus slew Phillip to revenge Marcus Annius Verus his father

and is the first Emperor in the Christian Era.so Thomas slew Lucius Arrias Verus to revenge James the Greater so Marcus Aurelius slew Thomas to revenge Emperor Lucius Arrias Verus his father. The remaining Apostle John Zebedee or John the evangelist did not join in the revenge of the priest. He was sentenced to life in prison and served in Pathos in Greece and died of old age.

Alternate Take Over the World Spell

Zacheus or Zachariah married Elizabeth and made John the Baptist.   Zacheus cheated and had Bernice, who was married to Herod Antipas, and so Herodias was born.   So John The Baptist liked Herodias and had her and was innocent because he did not know she was his half sister so his justice was done in the cell for incest by his Father because Incest is punishable by death by the Father in this case Zacheus.

So Alphy married Herodias and made Mathew. Then the Priestess the Virgin Mary Slaked Alphy and Made the lord, so Joachim determined who initiated the conception and killed his daughter for Incest.   So Mary Alphaeus slaked Alphy and Judas Thadeus and Mary Alphaeus was married to Cleopus, so she made Simon and Alphy took Mary Alphaeus and made James the Lesser and Barsabbas twins, so Hered Antipas, for cheating on his daughter slew Alphaeus.

So Judas Alpheus slew Herod Antipas, so Herod Phillip slew Judas Thadeus Brother Simon, so Judas Alpheus slew Herod Phillip, so Herod Agrippa slew his brother James the Lesser, so Herod Agrippa was slewed by Judas Alpheus.

So Judas Thadaeus had Bernice to revenge his brother James the Lesser who made Mariam Arrias who married Marcus Titus Flavius Sabinus so Bernice's husband Aristobulus the father of Perpetua who married Peter (Aristobulus) slew Judas Alphaeus so Andrew slew Aristobulus to revenge Judas Thadeus so Marcus Titus Flavius Sabinus slew Andrew so Peter slew Marcus Titus Flavius Sabinus to revenge Andrew so Gaius Sillius Calpurnius Domitius Piso slew Peter to revenge his father so Jonas slew Gaius Sillius Calpurnius Domitius Piso to revenge his son.

So Jonus had Mariame Caecina Arria Sabinus who made John Mark, so Arrius Antonius Calpernius Piso slew Jonas for the wife being raped So Peter's Mother Slew Arrius Antonius Calpernius Piso to revenge his father Jonas so Bonionia Prossilla Servila slew Peter's Mother to revenge her husband.

So Bartholomew slew Bonionia Prossilla Servilia the wife of Arrius Antonius Capernius Piso to revenge Peter so Arrius Antonius Calpernius Piso slew Bartholomew to revenge Bonionia Prossilla Servilia his wife so James the Greater slew Arrius Antonius Calpernius Piso to revenge Bartholomew, so Marcus Annius Verus slew James the Greater to revenge Arrius Antonius Calpernius Piso his father so Phillip did Battle and slew Marcus Annius Verus to revenge James the Greater so his son Lucius Arrias Verus slew Phillip to revenge Marcus Annius Verus his father and is the first Emperor in the Christian Era.so Thomas slew Lucius Arrias Verus to revenge James the Greater so Marcus Aurelius slew Thomas to revenge Emperor Lucius Arrias Verus his father. The remaining Apostle John Zebedee or John the evangelist did not join in the revenge of the priest. He was sentenced to life in prison and served in Pathos in Greece and died of old age.

Third Take Over the World Spell

First Star

Zacheus or Zachariah married Elizabeth and made John the Baptist.   Zacheus seduced and had Bernice, who was married to Herod Antipas, and so Herodias was born.   So John The Baptist liked Herodias and had her and was innocent because he did not know she was his half sister so his justice was done in the cell for incest by his Father because Incest is punishable by death by the Father in this case Zacheus.

Second Star

So Alphy slew Zacheus because he was married to Herodias and they made Mathew. Then the Priestess the Virgin Mary seduced and Slaked Alphy and Made the lord, so Joachim determined who initiated the conception and killed his daughter for Incest. So Alphy slew Joachim for killing his lover So Mary Alphaeus seduced and slaked Alphy and made James the Lesser and Barsabbas twins were born and Mary Alphaeus was married to Cleopus, so she made Simon with Cleopus and Alphy took Mary Alphaeus and Judas Thadeus was born and, so Hered Antipas, for cheating on his daughter slew Alphaeus.

Third Star

Judas Thadeus revenged his father Alphaeus and slew Herod Antipas, so Herod Phillip slew Judas Thadeus Brother Simon for revenge on his brother, so Judas Thadeus slew Herod Phillip, so Herod Agrippa slew his brother James the Lesser, so Herod Agrippa was slewed by Judas Thadeus.

Fourth Star

So Judas Thadaeus seduced Bernice to revenge his brother James the Lesser who made Mariam Arrias who married Marcus Titus Flavius Sabinus so Bernice's husband Aristobulus the father of Perpetua who married Peter, Aristobulus slew Judas Thadeus to revenge her seduction so Andrew slew Aristobulus to revenge Judas Thadeus so Marcus Titus Flavius Sabinus slew Andrew to revenge Aristobulus so Peter slew Marcus Titus Flavius Sabinus to revenge Andrew so Gaius

Sillius Calpurnius Domitius Piso slew Peter to revenge his father so To revenge his sons. Gaius Sillius Calpurnius Domitius Piso was slewed by Jonas.

So Jonus had Mariame Caecina Arria Sabinus who made John Mark, so Arrius Antonius Calpernius Piso slew Jonas for his wife being seduced So Peter's Mother Slew Arrius Antonius Calpernius Piso to revenge his father her husband Jonas so to revenge her husband Bonionia Prossilla Servila slew Peter's Mother wife of Jonas.

Fifth Star

So Bartholomew slew Bonionia Prossilla Servilia the wife of Arrius Antonius Capernius Piso to revenge Peter so Arrius Antonius Calpernius Piso slew Bartholomew to revenge Bonionia Prossilla Servilia his wife so James the Greater slew Arrius Antonius Calpernius Piso to revenge Bartholomew, so Marcus Annius Verus slew James the Greater to revenge Arrius Antonius Calpernius Piso his father so Phillip did Battle and slew Marcus Annius Verus to revenge James the Greater so his son Lucius Arrias Verus slew Phillip to revenge Marcus Annius Verus his father and is the first Emperor in the Christian Era so Thomas slew Lucius Arrias Verus to revenge James the Greater so Marcus Aurelius slew Thomas to revenge Emperor Lucius Arrias Verus his father. The remaining Apostle John Zebedee or John the evangelist did not join in the revenge of the priest. He was sentenced to life in prison and served in Pathos in Greece and died of old age.

Summaries

There are many ways to do this. Variance is found likely in the One Star General, Zacheus with or without innocence. In the Two Star Alphy revenges his sister, either by what he thinks is his father, or by his mother, but because they killed it is still law. Also the slake of Alphy by Mary wife of Cleopus to make the eldest son of Alphy and Mary, or if she had twins, or not or if Judas is older than James or not or if Simon is before or after the slake.

In the three Star there is variance about the order of sacrifice of the sons of Mary Alphaeus but in some texts Judas may kill only once, and then James and Simon could kill or not but likely James the lesser would be eldest to the slake to sacrifice the lord, revenged by his soldier brother Judas, and Simon the eldest of Cleopus also a sacrifice so of the sons of Mary Alphaeus would be the slaked lord the legitimate lord and finally the soldier lord and in the end Barsabbas becomes part of the loss in contest about the lot of the short straw while Mathias becomes the apostle, in the knight of Lots children.

There could be a way that involves Judas Iscariot but that puts it on the Priest. Likely what the lord did to take over the world was sealed by Judas Iscariot becoming a rabbi, the rapid heart beat when his mind discovers he has spilt his stomach out with a knife and there is no way for the temple of his body to survive.

The Survivors

The lord would have made this plan to save Mathew, because of the Gospel, and John Mark because of the Gospel, and John Zebedee because of the Gospel. In the Priest there is only one go up or grant of the blessing, so Jonas would be the sacrifice of the faith to make John Mark Peters Scribe.

Elizabeth, Cleopus, Mary Alphaeus, Perpetua, Bernice, Bernice, Mathew, John Mark. John the Evangelist

The Casualties

Zachariah, John the Baptist, Virgin Mary, Herodias, Alphaeus, Herod Antipas, James the Lesser, Herod Phillip, Judas Thadeus, Peter, Marcus Titus Flavius Sabinus, Andrew, Jonas, Gaius Sillius Calpurnius Domitius, Phillip. Arrius Antonius Calpernius Piso Barholomew, Arrius Antonius Calpernius Simon, Marcus Annius Verus, James the Greater.

House of David Side Casualties. 12

Levitical Priest Side Casualties, 9

Apostle Casualties   James the Lesser, Judas Thadeus, Peter, Andrew, Phillip, Bartholomew, Simon, James the greater, Thomas

Apostle Survivors   Mathew, John Zebedee

The Sins

First the Hebrew or House of David Side

Zachariah slept with another man's wife, who bore Herodias

John the Baptist had Herodias or the sin of incest

Alphaeus had sin of Matricide or Killing his mother and Adultery with Cleopus wife Mary

Anne Killed her daughter

Mary the Virgin Slaked

Judas Thadeus commited homicide to revenge his father

James the lesser commited homicide to revenge his brother

Simon committed adultery with another man's wife to make Miriam Arrias

Andrew Committed Homicide to revenge Simon

Jonas committed Homicide to revenge his eldest son

Jonus committed adultery to make John Mark the scribe of a gospel

Peter committed homicide to revenge his father

Bartholomew committed homicide of a wife to revenge Peter

James the Greater committed homicide to revenge Batholomew

Phillip committed homicide to revenge James the Greater

Thomas committed homice to revenge Phillip.

Second, the Levitical side of King Herod.

NOTE: The Levitical side did not actually destroy the criminal, but rather made a way to have each individual crucified at each post to control the world. They were all guilty of a mortal sin, except John the Evangelist who was given life in prison and Judas Iscariot who took his own life and is likely a Levitical Rabbi.

Herod Antipas committed homicide to revenge for a son in law committing adultery against his daughter

Herod Phillip committed homicide to revenge his brother

Herod Agrippa committed homicide to reveng his uncle

Marcus Titus Flavius Sabinus committed homicide to revenge the rape of his wife's mother

Gaius Sillius Calpurnius Domitius Piso commited homicide to revenge his father

Arrius Antonius Capernius Piso committed homicide to revenge his wife

Marcus Annius Verus committed homicide to revenge his father

Lucius Arrias Verus committed homicide to revenge his father

Marcus Aurelius committed homicide to revenge his father

The methodology of the derivation of the Take Over the World Spell

For each general or star, it must start with an us, Such as Zaccheus or Aphaeus, Thadeus, or Jonus, except the fifth star that is not part of us.

The blood shed must not kill the gospel books or the innocent.

The Apostles will take revenge on the Romans, who occupied the Holy Land, and justice must be maintained.

The order of the Roman people thru the making of each star, will be from King Herod and his line on down to Marcus Aurelius.

The faith is made, and the see of Rome is made, which opens to the see of Alexandria, or the idea of sharing Knowledge.

The sins are all atoned for, and the Christian Era is begun.

[26] Name the Stars.

(Generals)

Zacheus First Star, or Adjajjent (one us)

Alphaeus, who make Two Star, or Brigidier. (two us)

Thadeus, the three star or Major General (Three Us)

Jonus. The four star or General or (four us)

Not an us. And some Idiot put on five stars, which doesn't make any sense, so he can go to hell. So the Hellenism of Saint Helen makes cures out of the hell left by the commanding General.

[27]The Centurian Posts.

The Star of David is the one Star

The Centurian of Moses is the Two Star

The Centurian of Isaac is the Three Star

The Centurian of the Father Abraham is the Four Star

The Centurian of the Greek is Hercules or Helen is the Five Star

[28] During the War.

The Order of the Tribes is

Peter tribe of Simon Colonel and Naval Captain, heros Apollo and Ulysses

John tribe of Benjamin One Star hero Paris who kidnapped Helen

Bartholomew tribe of Dan Two Star and hero Perseus husband of Andromeda

James the Greater tribe of Judah Three Star and hero Persephone (wife of Hades)

Phillip tribe of Isaachar four Star, Hero Demetrius who married

Thomas tribe of Joseph Five Star Hero and God Dionysius

Order of Apostles during the War,

Order of Apostles

Mathew

Judas Thadeus

James the lesser

Simon

Andrew

Peter

Bartholomew

James the Greater

Phillip

Thomas

John the Evangelist

[29] War is Over

Takes place after Victory, and before peace is declared.

Five Star or Hercules is Retired or in disgrace, so Helen takes the Five Star to order the Cures made

Four Star is Reuben or Judas Iscariot

Three Star is Isaac-har or Phillip

Two Star is Levi or Mathew

One Star is Joseph or Thomas

Colonel is Benjamin or John the Evangelist

Lt. Colonel is Zebulon or James the Lesser

Major is Dan or Bartholomew

Captain is Asher or Simon

Lieutenant is Gad or Andrew

2nd Lieutenant is Naptoli or Judas Thadeus.

These are the ten tribes ordered by Ezekiel.

Both Judah and Simeon stay now in the Rear Admiral.

[30] Peace Time

Peace is declared and Victory is done, and the people return to life.

Judas Thadeus tribe of Naptoli Greek lady IO mother of Hero Epaphus rank 2nd Lieutenant

Mathew tribe of Levi Greek Lady Hera mother of Hero Hephaestus rank Lieutenant

Andrew tribe of Gad Greek Lady

James the Lesser tribe of Zebulon Greek Lady Deinara mother of Hero Hyllus rank Major

Simon tribe of Asher Greek Lady Ephrsus mother of Hero Minas rank Lt. Colonel

Peter tribe of Simon Greek Lady Leto mother of Hero God Apollo rank Colonel

John tribe of Benjamin Lady Hecuba mother of Hero Paris rank Adjujent Genera

Bartholomew tribe of Dan Greek Lady Danae mother of Hero Perseus rank Brigidier General

James the Greater tribe of Judah Greek Lady Demeter mother of Heroin Persephone rank Major General

Phillip tribe of Isaachar Greek Lady Rhea mother of Heroin Demeter rank General

Thomas tribe of Joseph Greek Lady Semele mother of Hero God Dionysius rank Commander General

Judas Iscariot tribe of Reuben Remaining Greek Ladies, and Heroes, and not a rank.

Apostle	Tribe	Greek Lady	Greek Hero	Rank
ANY	ANY	DRYOPE	PAN	LANCE CORPORAL
ANY	ANY	ALCEME	HERCULES	SARGEANT
JUDAS THADDEUS	NAPTALI	EPHRASUS	IO	2ND LEIUTENANT
MATHEW	LEVI	HERA	HEPHASUS	IST LEIUTENANT
ANDREW	GAD	MAIA	HERMES	CAPTAIN
JAMES THE LESSER	ZEBULON	DEINARIA	HYLLUS	MAJOR
SIMON	ASHER	LETO, EUROPA	MINOS	FIRST MATE, LIEUTENANT COLONEL
PETER	SIMEON	ANTICETUS, LETO	ULYSESS, APOLLO	CAPTAIN OF THE SHIP, COLONEL
JOHN	BENJAMIN	HECUBA	HECTOR	ADJUJANT GENERAL
BARTHOLOMEW	DAN	DANAE	PERSEUS	BRIGIDIER GENERAL
JAMES THE GREATER	JUDAH	DEMETER	PERSEPHONE	MAJOR GENERAL
PHILLIP	ISAAC-HAR	RHEA	DEMETER	GENERAL
THOMAS	JOSEPH	SEMELE	DIONYSIUS	COMMANDING GENERAL
JUDAS ISCARIOT	REUBEN	HARMONIA	ANY	HEBREW

[31] Apostles Veneration of Tribal Chieftains or the Spell of Making a Post

Apostle	Tribe Apostle Venerates	Veneration Number of Marriages	Marriages			Other Tribes
Judas Thaddeus	Naptali	2	Thalma wife of Neri	Anna wife of Abijah		All except Gad
Mathew	Levi	0	Mathew is the blood of Levi			All except Gad
Andrew of House Medes	Gad	Impossible				Unknown
James the Lesser	Zebulon	1	Barayah wife of Perez	Kanita wife of Hezron		All except Gad
Simon	Asher	Unknown 2 if Zibiah of Blood of Asher	Thalmar wife of Neri	Zibiah wife of Ahaziah of blood of Asher		All except Gad
Peter	Simon	Unknown				Unknown
John Zebedee	Benjamin	4	Salome; Joanne	Thalmar wife of Neri	Ahio of the line of Saul	All except Gad

Bartholomew	Dan	3	Thalmar wife of Neri	Ahio of the line of Saul	Hishim Granddaughter of Dan	All except Gad
James the Greater Zebedee	Judah	2	Salome married Judas Zebedee	JoAnne is male line of David		All except Gad
Phillip	Isaac-har	4	Miriam of line of King Herod;	Jehosebah daughter of Jerhoram; Athalia wife of Jefhoram	Ahijah wife of Nadab son of Jeraboam	All except Gad
Thomas	Joseph	2	Thalmar wife of Neri	Jecholiah wife of Amaziah or Athalia wife Queen of Jehoram		All except Gad
Judas Iscariot	Reuben	2	Jehosebah daughter of Jehoram	Rahab mother of Boaz of line of Reuben		Judah Zebulon Levi Isaac- har

Apostle	Literature	Country of Fate	Fate
Judas Thaddeus	Letter of Jude	Armenia	Martryed in Armenia

Mathew	Gospel of Mathew	Myrna Greece	Tradition is a Christian Literature the Martyrdom of Mathew In myrna by fire on a bed, with Bishop Plato
Andrew		Northwest of Black Sea	Crucified near Black Sea
James the Lesser	Letter of James	Israel	After Roman Govenor Festus Before Lucceaus Albinus A Great Fall From the Temple that injured and then stoned to death after a prophecy fullfillment in Jeremiah. Stoned by a Fuller
Simon		Israel	Brought before Roman Governor Atticus during Reign of Trajan and martyr in Holy Land
Peter	Gospel of Peter Letter of Peter	Rome Italy	Crucified in Nero's Garden
John	Gospel of John Gnostic book of John Revelations Letter of John 1 2 3	Greece Ephasus	Survived boiling oil, went to Patmos, Moved to Epheus, died at age of 94 in Epheus during Trajan
Bartholomew		Armenia	Flayed and beheaded by King Astyages in Armenia

James the Greater		Israel	Beheaded by King Herod Agrippa In Jerusalem
Phillip	Gospel of Phillip Apocalipse of Phillip	Greece	Crucified upside down during Emperor Domicletian in Greece
Thomas	Gospel of Thomas Gnostic book of Thomas	India	Crucified in India
Judas Iscariot	Gospel of Judas Iscariot	Israel	Suicide to make Rabbi in Potters Field
Heir to house of David in the Night Jesus Christ		Israel	Faked Crucifiction and lived to ripe old age

Strategic Positioning of Apostles during Disporia

It is known the the Lord of the Hebrews, Foster child of Saint Joseph in his second marriage or marriage of the night, was lord of most if not all of the kingdoms risen before the year one.

It is thought that the Lordship of the Christ was to unite all people in an effort to make a common government for the entire known world.

Consider the martyrdom of the apostles and lifes that they led after the crucifixtion of the lord and the sentencing of Nero in the Garden and how it was done.

[32] Idiot Spells

There are not parallel realities, there are people with sophisticated computer programs that brain wash people with computers remote operation, and televisions, using subliminal suggestion and other intelligence tricks. These people are dangerous idiots

Idi Amien ruler of Uganda, to discipline his prince son took his wife and cut off all limbs and switched left to right and right to left.

.

[33] Hand of War Spell

Father Bill said that a brother in the back of the truck kept his hand outside the truck.

Father driving almost hit the fence.

Brother hand was maimed, and he held it up and said I am truly blessed.

Reason for Kohath, son of Levi, the hand of war.

[34] Limp Dick Spell

Astarte Ningal Enkidu, NinMah, Anu, Anskar, Tiamet, Apophis, wife of Yam, Iluyanka

This spell is the difference about a Yank.   Possibly the origen of the Yank or Yankee term.   Made from the Sumatrian dieties. This spell will cause the victim to be completely hasseled and bothered with while a lady massages, sucks, or anything and the victim is continually distracted an cannot concentrate enough to get it up.

[35] Counter Limp Dick Spell

Astarte Ningal Enkidu, NinMah, Anu, Anskar, Tiamet, Apophis, wife of Yam, Iluyanka daughter Babys, Perecide son of daughter of Ion, Ion, Sol or Apollo.

To counter the Limp Dick Spell you must get away from Iluyanka. And go to the Greek God Apollo son of Zeus.

[36] Take a Dunk Man

Astarte Ningal Enkidu, NinMah, Anu, Enlil, Nergal, Enkidu, Ninmah, Adama

This spell is made out the sumatrian dieties, and will cause a man to deficate.

[37]Take a Dunk Women

Astarte Ningal Enkidu, NinMah, Anu, Enlil, Nergal, Enkidu, Nin-Khursag, Eve

Same as spell for Adamah, but is for Eve or Women.

[38] The Wind Spell

Astarte, Ningal, Enkidu, Ninmah, Anu God of the Sky, Enlil God of the Wind

This spell is a worship of the Jealous God Enlil. He is happy to make wind, sometimes in the strength of hurricanes or tempests.

[39] Spell of Division

When a deified God uses a hatchet to divide the black snake during the pregnancy of the mother subjected to more hel than any other women, who can assimilate the spirit of the serpent as well as the spirit of her noble father into the deified Gods Grandson, who is destined to make kingdoms topple and cure great diseases, stealing death from the powers that be and be more powerful than the possession the serpent uses, by possessing all gold and overcoming the pain of dispossession.

[40] Spells of Nursing

Rebecca Towne Nurse was an innocent lady, who was hung as a witch by Governor Phipps, in Salem Massachuesetts.   She was not a witch, she was not guilty of Killing Anyone.

Astarte, Ningal, Enkidu, Ninmah, Anu God of the Sky, Enlil God of the Wind, Ninlil the Beautiful Nurse

Rebecca had Twenty Four Cousins, including her cousin Deborah her nurse daughter of Huz.

[41] Jan Spell

Joseph as father of Janna the brother to his sister Janna who had Achim son of Zadok and Janna made Eliub who married Janna the daughter of Mathias and made Eleazor

[42] Spell of Doorway Lifting

Stand in a doorway.   Take both arms and press the back of your hands into the door Jam.   Use all your strength and hold for five minutes.

Step out of the Door, and relax your arms.   The spirit of the Doorway will lift your arms although not willed by the mind of the person who stood in the doorway.

[43] Spell to Defeat Asmoneus the Demon

Asomoneus was a levite that appartently turned evil.   He is mentioned in the book of Tobit, a sacred literature from the old testament were Azaziaz or The Archangel Raphael is mentioned as the person to defeat him.   Sarah had many husbands all killed by the demon Asamoneus.

Asamoneus is mentioned in the Anti Nicean father, a sect of Christianity that had arguments against what was declared true for the church by the council of Nicea,

Asmoneus father was the High Priest Onias the 47th High Priest.   Priests concerned in Asamoneus defeat is from Jaddua The 37th High Priest.

The priest must be completed in order from 37th, 38th, 39th etc.   It would be easy to do if the 38th priest was the father of the 39th and the 39th was father of the 40th etc,   this is not the case however.

Jaddua was the 37th high Priest.   The was succeeded by his son Onais the 38th high priest..   In the cool this is one the way. The 39th was the Simon the Just.   Now if Jaddua or Jade is the rock of royalty and his grandson is Sinon the just then everyone has to be treated justly. These are priests so that should go without says. The 40th high priest was the brother to Simon the Just, Eleazar. Eleazar was the 40th High Priest not difficult so far. Now 41 was the son of Jaddua Mannassah. Mannassah was succeeded by the son of Simon the Just. Now 42nd 43rd and 44th are direct sons of Simon the Just Onias the 2nd 42nd High Priest the father of Simon the 2nd the 43rd High Priest the father of Onias the 3rd the 44th High Priest. The 45th high priest was Jason the son of Simon the Just who is the 39th High Priest so you have to climb the lineage back to Simon the Just and down on generation to Jason High Priest. 46th High Priest was Menelaus the son of Simon the Just High Priest and brother to Jason. Now the 47th High Priest was Onias the 4th the father of Aaamoneus. So you see the problem in the succession of the High Priest.

In the Aaronic High Priest the succession was done by the most knowledgeable Priest to help the people. So it was not always from Father to son. The relationships in the succession of the High Priest would be very difficult to figure out. Thus would be difficult to follow.

Now Asamoneus is a direct ancestor of the Virgin Mary, and so also her sister Salome whose line continues to the King of Amorica Judiccal. So if a priest is angered all he has to do is put you in Asamoneus your ancestor and your Leviticus Is too difficult to figure out and basically you have to yield to the priest. Rafael the Arch Angel figured it out and now you can too.

Notice how many of the High Priest are easily shown to be our ancestors, however many like Jason the 45th High Priest are ancestors but not the same as most of the high priests because his son was not the High Priest at the time of his leaving the office, but a relative was.

Conclusions. In order to defeat Asamoneus you have to be able to climb 7 generations and father to children then climb back up to the father. Now the clencher.

The next High Priest, the 48th High Priest was not from the same line of Aaron like the house of the High Priest. He is Alcimus the 48th High Priest. The only requirement for a High Priest was to be descended from Aaron. The Levites usually stayed where they were all related. But Alcimus the next High Priest was an Aaronite however not from the lines like all the other high Priest.s In order to get to Alcimus then you have to go all the way back to Aaron and then down to Alcimus and to get to Jonathon Appus you have to go to Aaron Again and then to Mattathias the father of Johathon, Simon Thassi, and Judas Macceabeaus Elephant lord and great Warrior.

[44] Spells to Repel Demons for Mary Magdalene.

Now Mary Magdalene was possessed by seven demons according to the scriptures. These demons may be presented in the Gospel of Mary Magdalene that was found in the late 19th Century. These demons were darkness, Craving, ignorance, Lethal Jealousy, enslavement to the body, intoxicated wisdom, and guileful wisdom. There is a genealogy that explains the heresy and the truths and parallels that the heretical take advantage of.

The darkness curse is done with the iris of the eye. It can be judged to open or close to regulate the amount of light the subject sees. If closed enough the darkness can cause depression in the person due to the dim light.

Craving is difficult to counter, if about food possibly becoming a vegetarian, or if about sex, abstenance or satisfy the craving constantly until marriage or the lust is consumed.

The ignorance curse is the easiest to counter, however education or instruction takes a while.

Lethal Jealousy is difficult to contain, and if used will likely get someone killed. Not a good thing. Better to realize that if you really love someone that much then it is better to set them free because you love them.

Enslavement to the body is difficult, if its sex, ok, if its exercise, or if it's the idea you have to look good for your mate, then reason will likely repel this demon.

Intoxicated Wisdom is usually with alcohol or excessive work. To repel this demon either sober up or rest.

Guileful wisdom demon is repeled by humility, if you know you are not that smart then quit making arrangements for people.

[45] Spell to Resurrect the Dead

Dear Lord please resurrect the following person who has died (state the name) Post Christianity King Arthur who is blood of Heber Hyperion who marries Tamar the granddaughter of Josiah King of Judah who is a descendent of Jehosaphat a descendent of King David. Jehosaphat is married to Malika who is of the line of Gershom son of Levi. So from Malika navigate to Asaph who wrote many songs (straight up the line) who has many sons and is a boys choir, so the boys are the innocent and Michael the levite or Gershomite takes the dead soul from its carry and makes it part of the Pregnant Woman, whose child now has a soul that is both new and resurrected.

[46] Spell of Making the Lord

Heresy of Jesus Christ, the greatest Hebrew make
In order to make a king greater than David and Daniel then this king must be affiliated with Egyptians, Persians, Greeks, and the Roman Empire.

Now if Daniel is from two generations of Incest on both his mothers and fathers side and his parents were incestuous, then that much love would be hard to beat.   So the Roman Empire would be involved in creating a King of the Jews greater than King David, the father of Daniel, and yet also a Priest.

Now everyone knows that the Virgin Mary is the Mother of Jesus Christ.   Her only brother was Alphaeus.   Now we know that Ann was the mother of Mary, and also the mother of Alphaeus. Yet Daniel had two generations of Incest not one.   So Anne would have to marry her half brother to make the Virgin Mary and Alpheaus.   The Virgin would be born first and then Alphaeus.   Joachim would have Ann and make Salome, since his prayer ended up with Mary and the Temple.   So Anne's brother was Bar-Jesus the son of Bianca and Illegitimate, half brother to Anne who was also from Bianca.   Now Anne father was likely Joidiah so that Anne would be from the line of Aaron.   Bar Jesus father was the son of Bianca by Jeshua III, however has been called illegitimate from some sources. Now consider that Joseph of Arimathea was the brother to Bianca. So he is likely the father of Bar Jesus, not Jeshuah.   That's three generations of Incest that is better than Two, like Daniel. So the Hebrew have a greater make in Jesus Christ rather than King David.   King David is still the King over the jews, unless they go into Christianity.

The graphic shown below is likely part of a Heresy. It is certainly heretical and uses a different way of thought than what conventional Christians use.

So the make of a Lord is going to take a few generations. This is Similar to the make of David that also involves incest.

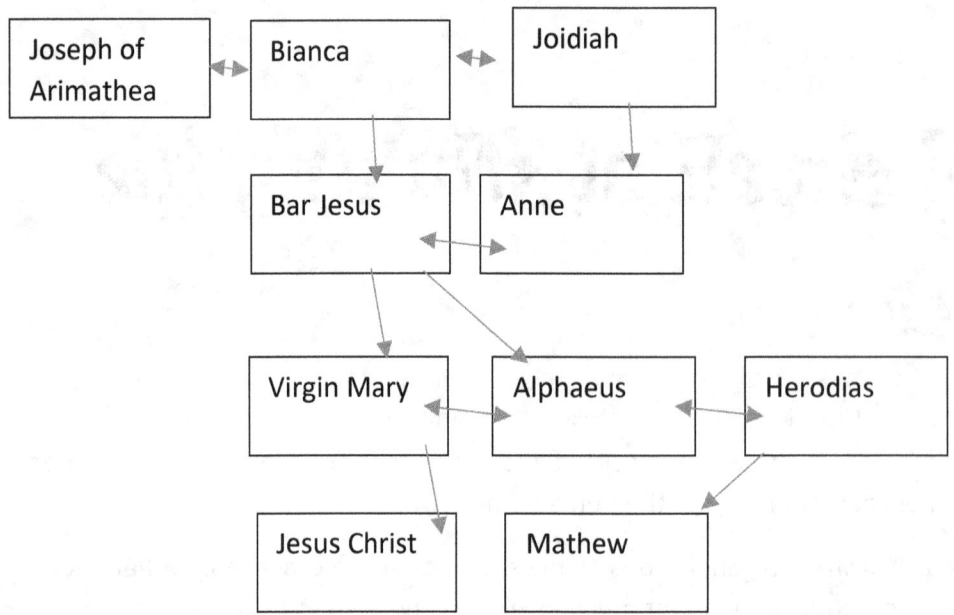

[47] Spell of Love Arimathea

The spell of Arimathea requires a Post of Saint Joseph of Arimathea and a correct veneration from Saint Caradoc of the Lord. These lines are hidden so that everyone who calls on the name of the Lord does not start a Arimathia. Also Medical doctors may prescribe heart medicine for an irregular heartbeat. Before Casting this it is a good Ideat to Post Saint Anne, and also make sure the lover is healthy enough. Also not for Children . Most of this is for the Mother and Father of whoever so that the Arimathea will generate enough love to last thru the night.

Jesus Christ, In the Resurrection

The graphic shown below is likely part of a Heresy. It is certainly heretical and uses a different way of thought than what conventional Christians use. Here the Enygenus marries the Lord and has a child Carodoc. For the Neophyte Carodoc is son of Bran (Bran was Duped) and Ann of Arimathea. The Resurrections of the Lord follow (Shown in Bold) so it eventually merges with the male line of Judah to form Strathcylde. It has to be presented as a heresy because the old and infirm and the children could be hurt if an arithmea is started.

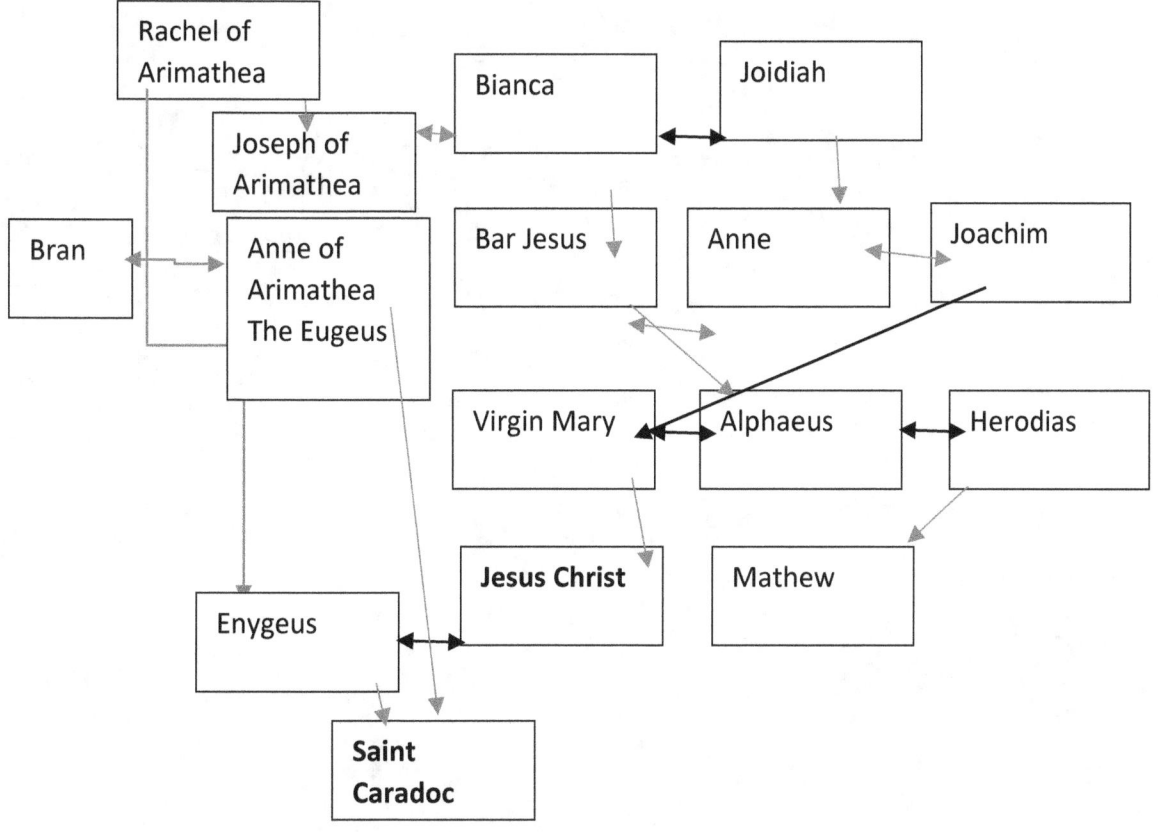

Navigate to Saint Anne of Arimathea.   Check the Euygeus and go to Rachel of Arimathea. If She wants you to feel love of Arimathea, she will post Anne the Priestess daughter of Joidiah the Levite and then post the true Eugeus from Caradoc and the line of Jesus the crucified son of Alphaeus son of Bar Jesus Son of Saint Joseph of Arimathea.

[48] Spell to Post the Greatest Christian King

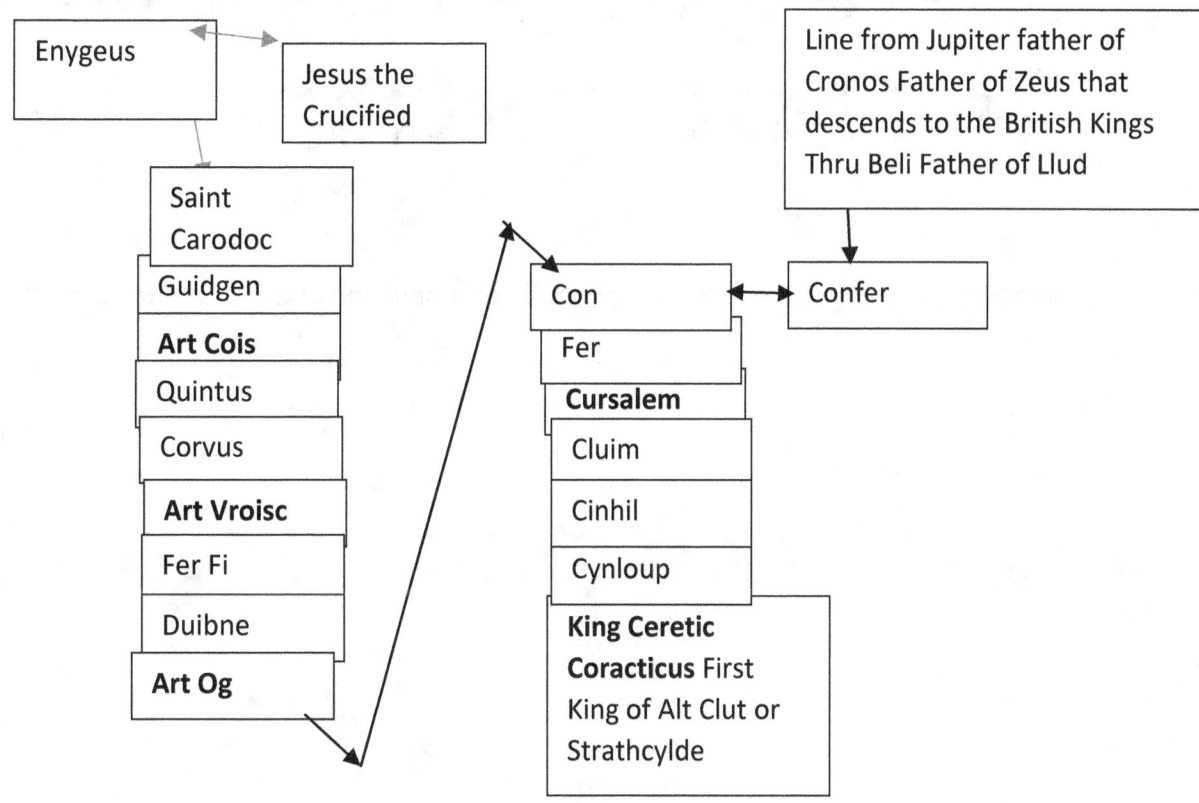

Saint Carodoc Grandson in the Resurrection Art Cois in the Resurrection Great Grandson Art Og in the resurrection CurSalem Add the mighty Arm of your Grandmother Line of Zeus to your High Priest Hilkiah.   Resurrection of CurSalem and the Lord with the mighty Arm of Zeus pull your Hilt and Cleve to fulfill the Law and Take Dumbarton Rock, King Ceretic Coracticus Rule your Davidic Kingdom with Justice and Law and give to all invaders their Tic of Slavery.

[49] Spell of Confusion

There is no invisible spell here.

[50] Spell To Throw Enemy To Chaos   [Greek]

Great God Zeus rise to your Father Cronos son of Uranus son of Gaia The Titanus to the Great Titan Chaos.

[51] Spell to Throw Enemy to Chaos [Israelie]

Judah son of Israel the way and the light rise to Israel and his wife the Daughter of the Sorceror and Israel rise to your Great Mother who is the faithful Intelligence and to the Child of the Promise Chaos (Isaac)

[52] Spell to Pass Thru Chaos and Create Voluptous Women

Great God Zeus rise to your Father Cronos son of Uranus son of Gaia The Titanus to the Great Titan Chaos Father of Eros Father of Erato who has Pschye Parents of Voluptua and the thought makes a Voluptus Women after many years.

[53] Spell to Throw Enemy to Chaos [Sumatrian]

Astarte Great Goddess your husband Cush Son of Ham Go to your Sacred Mother Emzara wife of Chaos (Noah)

[54] Spells To Rescue Oneself from Chaos

(Greek and Hebrew)   Chaos, Domnu , Descend to Dalbreach Father of Elathan Father of Ogma Father of Deleath and Bridgid mother of Donann who with Deleath made Brian or Orihah father of Kip whose line descends to Omar whose line descends to Heth Father of Uriah whose daughter marries the Dark Angel Uriel Angel of death whose daughter marries Rehoboam Grandson of King David.

[Sumatrian] Chaos, Father of Domnu marries The Granddaughter Of Olivia And Ham Father of Cush who has Astarte who has Dumuzi whose line Fathers Gordia who marries Cybele.   Great Goddess Astarte befuddle Cybele from her diabolical design and let the descent to Lydia continue to King Ardys of Lydia rise to Pluto daughter of Himas who marries the Daughter of Hades son of Cronos.

[55] Spell to make the Great King

Great Wizard Ibzan, son of Ruth and Salmon father of Edal Father of Abala cast to Jesse who posts his Gabe thru the Nether regions of the Galaxy to the Red Rays of Love that lead to the center of the make the Great King of Rigel and of the Milky Way David.

[56] War Spell

Mary Magdalene Mathias Simon Thassi, to post Miriam Maccabeus, to Mathias, to Judas Macabeus, move Ollie (First Macabees),   move Macy (Second Maccabeus),   Dumbo Awake (fourth Maccabeus)   keep the rhythm and the ryme and row (Third Maccabeus) and route the herd until they trample the enemy, row the lines, and when he slips, call Obidiah to Zerubabel, carry to Shazzrazah, to Nehemiah for the walls of Jerusalem to be rebuilt.

[57] Spell of Blessing

Judah, Bathshua, Hasadiah, Zedekiah, Masseiah, Neriah, Baruch (Baruch in this language at this time means Blessed so his name is Blessed)

[58] Spell to Raise the Sun

Astarte, Ningal, Enkidu, Ninmah, Anu, Anskar, Tiamet, Apophis, daughter of Yam, Iluyanka, daughter, Baby, Perecides, son of Ion, Ion, Sol, or Apollo

[59] Spell to Raise the Devil

Naptali, to Suri who has Sarvia the harlot and makes Joab whose daughter has Bichri father of Sheba.

Reuben to Hezron to Caleb to Hur and Miriam to Ezer and Miriam to Chelob, to Mehir to Tehenneh to Nahash and Abala to Sarvia.

Joseph the Israelite, Epriaim, descend to Azaziaz who has Sheba from the big Sur and makes Nebat who has the Harlot Sarvia, who makes the Devil himself.

[60] Spell to Raise Death

Jacobs son Levi, Kohath, Amram, Jocabed and Elizabeth mother of Nadab. (Death) with his brother Abihu.

[61] Spell to Raise Judgement

Ket Serug, Nahor, Terah, Abraham, Isaac, Israel, Dan or the divine Judgement

[62] Spell to Raise the Magician

Ket, Serug, Nahor, Terah, Abraham, Isaac, Israel, Joe the Magician and his brother Ben.

[63] Spell to Hang Ten Man

Ket, Serug, Nahor, Terah, Abraham, Isaac, Esau and Basemath, Timna, Amalek, Apag,Hammedatha,Haman father of the hanged, Dalphon, Poratha, Aspatha, Aridatha, Arisai, Adalia, Vajeezatha, Parshandatha, Parmashta, Aridah

[64] Spell to Raise the Staff

Ket, Serug, Nahor, Terah, Abraham, Isaac, Israel, Judah, Perez, Hezron, Jathir (Wood)

[65] Spell to Heighten For Fisticuffs.

King David, Abigail, Jeter, Jada, Onam, Jeremeel, Hezron (Right Cross)

[66] Spell to Make the Martialis win the contest

Jesse, Abigail, Daniel, to Hilkiah, to Gamaliel post Simon father of Imme Shalom, and From Mathan, Hizkiah, Judas the Zealot, Simon, Hyrancus marry the Imme Shalom, descend Eleazar Ha Gadol, Simon, Gamaliel, Hillel, Malchion, son, Conaus the Martyr, son Salomus (Shallom, Salomon) Urbanus, Mansuetus, Symetrius, Martialis,------- Strike

The second blow is an insult to the Saint, because the first Blow is Fatal

[67] Spell In the Knight to Add Food and Shelter to the Pregnant Women with Child

Mathan, Jacob, Eucheria, Miriam, Theudas, Addai James, Unknown Father of Sotor,

Unknown Selius goes to Ptolas or Bartelmaeus, and buys the beasts in Worship of the Man who proved he was god of that beast, and then to Ham and Olivia to Domnu to Add the Songs of Life to the child and then to Paratama the Overlord, and to Shaushatar the wood Lord and to Aratama the Horse Lord, and to Shuttarna the War Lord and to Tushratta the Axe Lord.

[68] Spell to Cure Arthritus

Saint, Saint, Saint, Joe of Arthimea in the right to Saint Joseph of Ariamathea

[69] Spell to Open

the ear when the skull closes on it

When the sin of the father is so bad those in judgement cause the bones of the skull to close the ear canal, then this spell will open it and allow the Priest or Rabbi to see the sins of the subjects father.

Ket, Serug, Nahor, Terah, Abraham, Isaac, Israel, Rueben, Hezron, Caleb, Hur, Epratha

Ket, Serug, Nahor, Terah, Abraham, Isaac, Israel, Joseph, Mannassah, Machir, Epratha.

[70] Spell to Cure Cancer

Oh Great King and Saint Ethelbert Please cure this Cancer.

Saint Ethelbert believe in God Odin and ask the favor of God's Wife Frigg whose lineage marries Buonduah whose lineage marries Ann the Daughter of Jose the Rama Thea the son of the Saint Joseph of the Davidic Kingdom.   Saint Joseph post your daughter Ann and ask your grandfather Mathan favor with his daughter Don who marries Mathan Great Grandson Beli.

[71] Spell to Raise the Tempest

Charlemagne   Great emperor *Karolus serenissimus augustus a Deo coronatus magnus, pacificus imperator, Romanum gubernans imperium, qui et per misericordiam Die rex Francorum atque Langobardorum* let your line descend to the Vermandois to the Brittany to King William Conquer !!!!!   King Henry of England, Meshenines, Tempest Rise and consume the enemy.

[72] Take Over the World Spell thru the Son

House of David, post the formation of Christianity and the feat of Don marries Beli, 4 Generations

Post the Big Sur the feat of Sarvia mothers Jeroboam, 5 generations

Post the Healing Angel Rafael whose sister Sarah Marries Tobias 6 Generartions

Finally increase the Son Worship where Rahotep Grandson of Pepi marries the ninth generation daughter of Yakabom   7 Generations.

Long Live Ra the Sun

Section Two Spells Of Magick

[2] Itch spell

Daniel is father of Jacob who is father of Joseph who is father of Mannasseh who is father of Ebenezer who is father of Moses who is father of the prophet Malachi……

Incidently Joseph has an angel.

This is a part of a line that goes from Daniel Son of Abigail, to Saint Gamaliel.   It is watched due to the mention of Moses.

[4] Death Spell

Mene, Mene, Tekel, Perez. (Mene, Mene, Shekel Phares.)

If you can't fart, and you can't burp, then in three days the pig inside you causes internal bleeding.

A Ghastly way to die. (50 Shekels are paper bills)

This spell is found in the Hebrew book of Daniel.   It is cast by god and Daniel's keeper Beltzasharhar dies.

The mene parts are full menestrations by women, the price is the Shekel, a type of coin and bill the Hebrew use and the Phares is like the Pharoah or the priest.

[5] Dumb Spell

Causes dumbness or the inability to speak in the presence of a beautiful women

[6] Astarte had Zabada who made Shala who married Enlil (God) and then Shala had Iskur who made Dumuzi who had Astarte and made a daughter who had Gilgamesh.

This is a navigation from Astarte, who married Cush, his son Nimrod, and his son Gilgamesh, and Zabada and Dumuzi.

[6] Luck Spell

King Arthur line continued to Heber Hyperion who married Tamar the granddaughter of King Josiah, whose wife was the daughter of Pediah who was the son of Jecoliah who was the son of Jehoikim who married Nahusta of the line of Gad that married the line of Sharmariah Son of David.

This is the Shamrock spell, or veneration of the Son of King David.   Shamariah has descendents that marry the tribe of Gad. However at King Josiah, a Hebrew king, the ascent goes to King Joasiah, and then to his other wife, so this is a navigation not a veneration.

Strength Luck,   Mathan, Hizkiah, Judas the Zealot, Simon, Eleazar, Andrew Lukuas

[7] Health Spell

Jacob is father of Naptoli who is father of Asiel who is father of the Angel Raguel who is Father of the Archangel of healing Rafael and his sister Sarah.

Rafael is the healing angel, and is an archangel.   Jacob is Israel or the father of twelve Israelites and Naptoli is one Israelite.

[8] Health Spell utilizing a dying person.

[9] Rafael is father of Gabael who is father of Aduel who is father of Hananiel who marries Deborah of Epraim who is mother of Tobiel who is father of Tobit who marries Anna who is father of Tobias who marries Sarah the brother of Rafael.

This is similar to the last spell, except the line continues. Notice Sarah is six generations Tobias Grandmother.

[9] The Wheel spell (involves the Astarte Goddess Tara.)

TAROTAROTAROTARA

The Wheel in the above spell will post three O and then the most intelligent mind of Astarte Goddesses will interpret. Can be used for all commissioned rank, depending on how many ties the wheel goes around. For each Tarot is the first, second, third commissioned rank depending on how much Tara knows about eact rank. Possible to go to eleven because there are only eleven commissioned Ranks. O-12 is reserved for the Lord.

O-1 One time around the wheel, Tara is like the lady for the second lieutenant

O-2 Two time around the whell, Tara is the lady who likes the first lieutenant

O-3 Tara Likes the Captain

O-4 Tara like the Major

O-5 Tara likes the Lt. Colonel

O-6 Tara like the Colonel of if Navy the Captain of the Ship

O-7 Tara likes the Adjunct General 1-star

O-8 Tara likes the Brigidier General 2-star

O-9 Tara likes the Major General   3-star

O-10 Tara likes the General   4-star

O-11 Tara likes the Commanding General 5-star

The Tarot has always been associated with Tara, who is usually gifted as a medium or astrologist if she decides to use this talent.

[10] The Earthquake Spell.

(spoken by Phillip at his crucifixtion)

And he began to curse them, invoking, and crying out in Hebrew: Abalo, aremun, iduthael, tharseleon, nachoth, aidunaph, teletoloi: that is, O Father of Christ, the only and Almighty God; O God, whom all ages dread, powerful and impartial Judge, whose name is in Thy dynasty Sabaoth, blessed art Thou for everlasting: before Thee tremble dominions and powers of the celestials, and the fire-breathing threats of the cherubic living ones; the King, holy in majesty, whose name came upon the wild beasts of the desert, and they were tamed, and praised Thee with a rational voice; who lookest upon us, and readily grantest our requests; who knewest us before we were fashioned; the Overseer of all: now, I pray, let the great Hades open its mouth; let the great abyss swallow up these the ungodly, who have not been willing to receive the word of truth in this city. So let it be, Sabaoth. And, behold, suddenly the abyss was opened, and the whole of the place in which the proconsul was sitting was swallowed up, and the whole of the temple, and the viper which they worshipped, and great crowds, and the priests of the viper, about seven thousand men,

The Church Fathers. The Complete Ante-Nicene & Nicene and Post-Nicene Church Fathers Collection: 3 Series, 37 Volumes, 65 Authors, 1,000 Books, 18,000 Chapters, 16 Million Words (Kindle Locations 162425-162429). Catholic Way Publishing. Kindle Edition.

The Church Fathers. The Complete Ante-Nicene & Nicene and Post-Nicene Church Fathers Collection: 3 Series, 37 Volumes, 65 Authors, 1,000 Books, 18,000 Chapters, 16 Million Words (Kindle Locations 162421-162425). Catholic Way Publishing. Kindle Edition. Ephratha ? ?

This is a dangerous spell and is cast when Phillip is undergoing public execution. The moral of the story here is do not go to a public execution.

[11] The spell of Ishmael

(made in the seventh month of a pregnancy, so she will make milk for her child) [Part of Gabriel]

Gabel to Penuil   break Ishmael.

There is a line in every man and woman that connects the tits, nobs, breast, gabes, etc.. to the hip. This connection is patrolled and if making Gabriel it is Gabriel.   Ishmael is mothers milk but it takes a few months for the glands to transition to be suppliers of felicity to the baby.

[12] The Love Spell or Shine Spell

Bathsheba, David, please allow your love from generations of incest to channel to me and shine like a diamond.   Your parents loved each other since birth, and thru love making, and each generation makes more shine.   If you have abundant love I know sometimes you wish to share. Thank you.

[13] Spell to Sex Child to Preference

Lud Law to Caesars to the Caesar Flavius Sextus who is a descendent of Flavius Julius Augustus Caesar.

This spell is a veneration, keep in mind that Flavius Sextus is not the chosen line of the Emperors, but is a second line.

[14] Spell to Make living thing (can neither be destroyed or Altered) (Dominic)

[15] Songs of Life Spells

Olive had Ham whose child was the parent of the Wife of Domnu the son of Chaos and Nxthy the children of Caligo and Athys.

The song of Domnu is the song of life for all things God had made.

The Olive tree starts with car------

The lamb Domnu starts with Michaiah........

The Corn Plant Domnu starts with Whitigen.........

The Snake Donmu starts with Cleoputre............

The Hela Monster Donmu starts with Gihon..............

[16] The spell of increasing the Psychyey or Psycopath or keep the children on the right Path.

Caligo had Athys who had Chaos Who had Nytx who had Erebos who had Nytx make Eros who had Psyche who made Voluptua.   (Voluptus Daughter)

Great God Zeus let your titan father Cronus remember his father Uranus whose mother was Gaia who married Chaos Who had Nytx who made Erebos who had Nytx make Eros who had Psyche who made Voluptua.   (Voluptus Daughter)

[17] The metallurgic spell of making Brass

Copper and Bronze make Brass at the forge.

[18] Control the speed of Time

Ithreel is the time stream   King Saul is father of Eglah who had Phatiel the son of Laish the son of King Saul.   Eglah also had King David who was the father of Ithream.

This spell is part of King David. It can be dressed up with Jupiter and Saturnus, for forward and backward travel in the Ithream time stream monitors by the King.

[19] The spell of fire (male and female)

Caligo had Athys who made Chaos who had Nytx who made Domnu whose line continued to DelBreath whose line continued to Donan who was mother of Brian, Iachar, and Iacharba.

Iachar, and Iacharba are fire man and fire women. Char makes fire.   Ia is like I am. Navigate to Chaos and cast away !!!

[20] Become a Wizard and Raise the Angel of Death

Ruth had Salomon and made Ibzan who was the father of Edal who was the Father of Abala who had Jesse but also Uriel the dark Angel   (Restores youth thru labor)

This is not known since Ruth is married to Boaz and Salomon is Boaz father.   So Ibzan is the wizard, and the line continues to Abala who married Jesse of Bethlehem, but slept with the levite Uriel who is the angel of death or the dark angel.

[21] Cure for Appendicitis

Astarte had Nimrod who had a son who had Jared the Father of Oriah the father of Kip whose line goes to Omar whose line goes to Heth, the father of Uriah, the father of Absalom the father of Anna who has Rehoboam and makes Abijah

[22] Death Spell of Broken Neck or Make an Angel

Aaron is father of Ithamar is father of Eli.

This is commonly done when Samuel is read by a person.   The subject is made to fall, and his neck bone breaks.   So the doctor puts the head in a halo, or brace so is neck does not move. This spell may cause death, and healing the bone takes about three months, so the person becomes an angel with a halo for three months.

[23] Counter Death Spell of Broken Neck

Aaron is father of Ithamar is father of Eli is father of a daughter (Crystal, or Christina, or Chrissy) who marries the high priest Uzzi and is also named Ely.

This spell is used because Eli is the Levite priest that makes the fall, however has a daughter that is not really well known and so marries the Levite Uzzi and gets the person who receives the Death Spell into a different Priest.

[24] Save from Death from Israel or Keep the death in Israel

The Jew-is

The 6th King of Judah Jehoram married the 7th King of Israel Ahab's daughter Athalia the 8th queen regent of Judah whose Brother Jehoram was the 9th King of Israel

The history is the Israeli King Jehu comes after Jehoram.   So the priest makes an attack on the subjects Judan Jehoram, however is countered by this spell and the Jehu harvest is done to Jehoram the King of Israel not Jehoram the King of Judah.   Then the covenant between Jesse of Bethlehem and Jeroboam is invoked that is that Jeroboam, and his descendents, will shepard and not kill Solomon's descendents, who are Kings of Judah.

[25] The Take Over the World Spell

The Tear or the Generals   Five Stars

Zacheus or Zachariah married Elizabeth and made John the Baptist.   Zacheus cheated and had Bernice, who was married to Aristobulos, and so Herodias was born.   So John The Baptist liked Herodias and had her so his justice was done in the cell for incest

So Zacheus was sacrificed in the doorway by Alphy to revenge his wife's rape.

So Alphaeus married Herodias, who had Mathew, so the Virgin Alphy's half sister slaked Alphy and she made the lord Jesus, so Ann slew the Virgin for slaking and that was her Justice for Adulterous Incest. So Alpheus slew his mother for killing his lover.

So Alphaeus had Cleopus wife Mary because she slaked him and Mary made James the lesser and Judas Thadeus. So Herod Antipas slew Alphaeus for cheating on his daughter.

So Judas Thadeus revenged his father and slew Herod Antipas so the blood of King Herod was split so Herod Phillip slew Judas Thadeus to revenge his brother, so his brother James the Lesser slew Herod Phillip, so Herod Agrippa II slew James the Lesser to revenge his Uncle, so Herod Agrippa II had Bernice who made Perpetua.

Herod Agrppa II was married to Bernice But Simon had Bernice to revenge his brother James the Lesser who made Mariam Arrias who married Marcus Titus Flavius Sabinus so he slew Simon to revenge the rape of his wife's mother so Andrew slew Marcus Titus Flavius Sabinus to revenge Simon so Gaius Sillius Calpurnius Domitius Piso slew Andrew to revenge his father so Jonas slew Gaius Sillius Caopurnius Domitius Piso to revenge his eldest son.

So Jonus had Mariame Caecina Arria Sabinus who made John Mark, so Arrius Antonius Calpernius Piso slew Jonas for the wife being raped So Peter Slew Arrius Antonius Calpernius Piso to revenge his father Jonas so Bonionia Prossilla Servila slew Peter to revenge her husband.

So Bartholomew slew Bonionia Prossilla Servilia the wife of Arrius Antonius Capernius Piso to revenge Peter so Arrius Antonius Calpernius Piso slew Bartholomew to revenge Bonionia Prossilla Servilia his wife so James the Greater slew Arrius Antonius Calpernius Piso to revenge Bartholomew, so Marcus Annius Verus slew James the Greater to revenge Arrius Antonius Calpernius Piso his father so Phillip did Battle and slew Marcus Annius Verus to revenge James the Greater so his son Lucius Arrias Verus slew Phillip to revenge Marcus Annius Verus his father

and is the first Emperor in the Christian Era.so Thomas slew Lucius Arrias Verus to revenge James the Greater so Marcus Aurelius slew Thomas to revenge Emperor Lucius Arrias Verus his father. The remaining Apostle John Zebedee or John the evangelist did not join in the revenge of the priest. He was sentenced to life in prison and served in Pathos in Greece and died of old age.

Alternate Take Over the World Spell

Zacheus or Zachariah married Elizabeth and made John the Baptist.   Zacheus cheated and had Bernice, who was married tp Aristolbulos, and so Herodias was born.   So John The Baptist liked Herodias and had her and was innocent because he did not know she was his half sister so his justice was done in the cell for incest by her Father Aristobulos   So Aristobulos Slew Zacheus for cheating on his wife.

So Alphy married Herodias and made Mathew. Then the Priestess the Virgin Mary Slaked Alphy and Made the lord, so Joachim determined who initiated the conception and killed his daughter for Incest. So Mary Alphaeus slaked Alphy and Judas Thadeus and Mary Alphaeus was married to Cleopus, so she made Simon and Alphy took Mary Alphaeus and made James the Lesser and Barsabbas twins, so Hered Antipas, for cheating on his daughter slew Alphaeus.

So Judas Alpheus slew Herod Antipas, so Herod Phillip slew Judas Thadeus, so Simon slew Herod Phillip, so Herod Agrippa slew Simon, so Herod Agrippa was slewed by James the Lesser.

So James the Lesser had Bernice to revenge his brother Simon who made Mariam Arrias who married Marcus Titus Flavius Sabinus so Bernice's husband Aristobulus the father of Perpetua who married Peter (Aristobulus) slew James the Lesser so Andrew slew Aristobulus to revenge James the Lesser so Marcus Titus Flavius Sabinus slew Andrew so Peter slew Marcus Titus Flavius Sabinus to revenge Andrew so Gaius Sillius Calpurnius Domitius Piso slew Peter to revenge his father so Jonas slew Gaius Sillius Calpurnius Domitius Piso to revenge his son.

So Jonus had Mariame Caecina Arria Sabinus who made John Mark, so Arrius Antonius Calpernius Piso slew Jonas for the wife being raped So Peter's Mother Slew Arrius Antonius Calpernius Piso to revenge his father Jonas so Bonionia Prossilla Servila slew Peter's Mother to revenge her husband.

So Bartholomew slew Bonionia Prossilla Servilia the wife of Arrius Antonius Capernius Piso to revenge Peter so Arrius Antonius Calpernius Piso slew Bartholomew to revenge Bonionia Prossilla Servilia his wife so James the Greater slew Arrius Antonius Calpernius Piso to revenge Bartholomew, so Marcus Annius Verus slew James the Greater to revenge Arrius Antonius Calpernius Piso his father so Phillip did Battle and slew Marcus Annius Verus to revenge James the Greater so his son Lucius Arrias Verus slew Phillip to revenge Marcus Annius Verus his father

and is the first Emperor in the Christian Era.so Thomas slew Lucius Arrias Verus to revenge James the Greater so Marcus Aurelius slew Thomas to revenge Emperor Lucius Arrias Verus his father. The remaining Apostle John Zebedee or John the evangelist did not join in the revenge of the priest. He was sentenced to life in prison and served in Pathos in Greece and died of old age.

Third Take Over the World Spell

First Star

Zacheus or Zachariah married Elizabeth and made John the Baptist.   Zacheus seduced and had Bernice, who was married to Aristobulos, and so Herodias was born.   So John The Baptist liked Herodias and had her and was innocent because he did not know she was his half sister so his justice was done in the cell for incest by his Father because Incest is punishable by death by the Father in this case Zacheus.

Second Star

So Aristobulos slew Zacheus because his wife was raped by Zacheus. So Alphy had Herodias and they made Mathew. Then the Priestess the Virgin Mary seduced and Slaked Alphy and Made the lord, so Joachim determined who initiated the conception and killed his daughter for Incest. So Alphy slew Joachim for killing his lover So Mary Alphaeus seduced and slaked Alphy and made James the Lesser and Barsabbas twins were born and Mary Alphaeus was married to Cleopus, so she made Simon with Cleopus and Alphy took Mary Alphaeus and Judas Thadeus was born and, so Hered Antipas, for cheating on his daughter slew Alphaeus.

Third Star

Judas Thadeus revenged his father Alphaeus and slew Herod Antipas, so Herod Phillip slew Judas Thadeus for revenge on his brother, so Simon slew Herod Phillip, so Herod Agrippa slew Simeon, so Herod Agrippa was slewed by James the Lesser.

Fourth Star

So James the Lesser seduced Bernice to revenge his brothers Simon and Judas Thadeus who made Mariam Arrias who married Marcus Titus Flavius Sabinus so Bernice's husband Aristobulus the father of Perpetua who married Peter, Aristobulus slew James the Lesser to revenge her seduction so Andrew slew Aristobulus to revenge Judas Thadeus so Marcus Titus Flavius Sabinus slew Andrew to revenge Aristobulus so Peter slew Marcus Titus Flavius Sabinus to revenge Andrew so Gaius Sillius Calpurnius Domitius Piso slew Peter to revenge his father so To revenge his sons. Gaius Sillius Calpurnius Domitius Piso was slewed by Jonas.

So Jonus had Mariame Caecina Arria Sabinus who made John Mark, so Arrius Antonius Calpernius Piso   slew Jonas for his wife being seduced So Peter's Mother Slew Arrius Antonius Calpernius Piso to revenge his father her husband Jonas so to revenge her husband Bonionia Prossilla Servila slew Peter's Mother wife of Jonas.

Fifth Star

So Bartholomew slew Bonionia Prossilla Servilia the wife of Arrius Antonius Capernius Piso to revenge Peter so Arrius Antonius Calpernius Piso slew Bartholomew to revenge Bonionia Prossilla Servilia his wife so James the Greater slew Arrius Antonius Calpernius Piso to revenge Bartholomew, so Marcus Annius Verus slew James the Greater to revenge Arrius Antonius Calpernius Piso his father so Phillip did Battle and slew Marcus Annius Verus to revenge James the Greater so his son Lucius Arrias Verus slew Phillip to revenge Marcus Annius Verus his father and is the first Emperor in the Christian Era so Thomas slew Lucius Arrias Verus to revenge James the Greater so Marcus Aurelius slew Thomas to revenge Emperor Lucius Arrias Verus his father. The remaining Apostle John Zebedee or John the evangelist did not join in the revenge of the priest. He was sentenced to life in prison and served in Pathos in Greece and died of old age.

Fourth Take Over the World Spell

First Star

Zacheus or Zachariah married Elizabeth and made John the Baptist.   Zacheus seduced and had Bernice, who was married to Aristobulos, and so Herodias was born and Zacheus lost his soul for adultery and covet of his neighbors wife.   So John The Baptist liked Herodias and had her and was innocent because he did not know she was his half sister so his justice was done in the cell for incest by his Father because Incest is punishable by death by the Father in this case Zacheus and John the Baptist died in innocence

Second Star

So Alphy was married to Herodias and they made Mathew and Alphy slew Zacheus to revenge his mother in law's rape.   Then the Priestess the Virgin Mary seduced and Slaked Alphy and Made thru an immaculate conception the lord, so Joachim the Virgin's Father had Ann and made Salome, and then determined who initiated the conception and Joachim killed his daughter for Incest. So Alphy slew Joachim for killing his lover So Mary Alphaeus seduced and slaked Alphy and so lost her soul and then made thru an immaculate conception James the Lesser and Barsabbas who were twins. So Mary Alphaeus was married to Cleopus, so she made Simon with Cleopus and Alphy took Mary Alphaeus and so lost his soul thru adultery and covet of his neighbors wife and Judas Thadeus was born and, so Hered Antipas, for cheating on his daughter slew Alphaeus and Bar Jesus slew Mary Alphaeus for slaking Alphy.

Third Star

Judas Thadeus revenged his father Alphaeus and slew Herod Antipas, so Herod Phillip slew Judas Thadeus for revenge on his brother, so Simon slew Herod Phillip, so Herod Agrippa slew Simon, so Herod Agrippa was slewed by Simon's Father Cleopus.

Fourth Star

So James the Lesser seduced Bernice to revenge his brothers Simon and Judas Thadeus who made Mariam Arrias who married Marcus Titus Flavius Sabinus so Bernice's husband Aristobulus the father of Perpetua who married Peter, Aristobulus slew James the Lesser to revenge her seduction so Peter slew Aristobulus his father in law to revenge Judas Thadeus so Marcus Titus Flavius Sabinus slew Peter to revenge Aristobulus so Andrew slew Marcus Titus Flavius Sabinus to revenge Peter so Gaius Sillius Calpurnius Domitius Piso slew Andrew to revenge his father so To revenge his sons. Gaius Sillius Calpurnius Domitius Piso was slewed by Jonas.

So Jonus had Mariame Caecina Arria Sabinus who made John Mark, so Arrius Antonius Calpernius Piso slew Jonas for his wife being seduced So Peter's Mother Slew Arrius Antonius Calpernius Piso to revenge his father her husband Jonas so to revenge her husband Bonionia Prossilla Servila slew Peter's Mother wife of Jonas.

Fifth Star

So Bartholomew slew Bonionia Prossilla Servilia the wife of Arrius Antonius Capernius Piso to revenge Peter so Arrius Antonius Calpernius Piso slew Bartholomew to revenge Bonionia Prossilla Servilia his wife so James the Greater slew Arrius Antonius Calpernius Piso to revenge Bartholomew, so Marcus Annius Verus slew James the Greater to revenge Arrius Antonius Calpernius Piso his father so Phillip did Battle and slew Marcus Annius Verus to revenge James the Greater so his son Lucius Arrias Verus slew Phillip to revenge Marcus Annius Verus his father so Thomas slew Lucius Arrias Verus to revenge James the Greater so Marcus Aurelius slew Thomas to revenge Emperor Lucius Arrias Verus his father. The remaining Apostle John Zebedee or John the evangelist did not join in the revenge of the priest. He was sentenced to life in prison and served in Pathos in Greece and died of old age.

Fifth Take Over the World Spell

First Star

Zacheus or Zachariah married Elizabeth and made John the Baptist. Zacheus seduced and had Bernice, who was married to Aristobulos so Herodias was born and Zacheus lost his soul for adultery and covet of his neighbors wife. So John The Baptist liked Herodias and had her and was innocent because he did not know she was his half sister so his justice was done in the cell for incest by his Father because Incest is punishable by death by the Father in this case Zacheus and John the Baptist died in innocence

Second Star

So Alphy was married to Herodias and they made Mathew and Herod Antipas slew Zachariah in the doorway to revenge his wife's mothers's rape. Then the Priestess the Virgin Mary seduced and Slaked Alphy and Made thru an immaculate conception the lord, so Bar Jesus Alphy's Father had Ann and made Salome, and then determined who initiated the conception and Bar Jesus killed the Virgin Mary for Incest. So Mary Alphaeus was married to Cleopus, so she made Simon with Cleopus So Mary Alphaeus seduced and slaked Alphy and so lost her soul and then made thru an immaculate conception James the Lesser and Barsabbas who were twins. and Alphy took Mary Alphaeus and so lost his soul thru adultery and covet of his neighbors wife and Judas Thadeus was born. So Bar Jesus slew Mary Alphaeus for slaking his son and, so Hered Antipas, for cheating on his daughter slew Alphaeus.

Third Star

Judas Thadeus revenged his father Alphaeus and slew Herod Antipas, so Herod Phillip slew Judas Thadeus Brother Simon for revenge on his brother, so Judas Thadeus slew Herod Phillip, so Herod Agrippa slew his brother James the Lesser, so Herod Agrippa was slewed by Judas Thadeus.

Fourth Star

So Judas Thadaeus seduced Bernice to revenge his brother James the Lesser who made Mariam Arrias who married Marcus Titus Flavius Sabinus so Bernice's husband Aristobulus the father of Perpetua who married Peter, Aristobulus slew Judas Thadeus to revenge her seduction so Peter slew Aristobulus his father in law to revenge Judas Thadeus so Marcus Titus Flavius Sabinus slew Peter to revenge Aristobulus so Andrew slew Marcus Titus Flavius Sabinus to revenge Peter so Gaius Sillius Calpurnius Domitius Piso slew Andrew to revenge his father so To revenge his sons. Gaius Sillius Calpurnius Domitius Piso was slewed by Jonas.

So Jonus had Mariame Caecina Arria Sabinus who made John Mark, so Arrius Antonius Calpernius Piso slew Jonas for his wife being seduced So Peter's Mother Slew Arrius Antonius Calpernius Piso to revenge his father her husband Jonas so to revenge her husband Bonionia Prossilla Servila slew Peter's Mother wife of Jonas.

Fifth Star

So Bartholomew slew Bonionia Prossilla Servilia the wife of Arrius Antonius Capernius Piso to revenge Peter so Arrius Antonius Calpernius Piso slew Bartholomew to revenge Bonionia Prossilla Servilia his wife so James the Greater slew Arrius Antonius Calpernius Piso to revenge

Bartholomew, so Marcus Annius Verus slew James the Greater to revenge Arrius Antonius Calpernius Piso his father so Phillip did Battle and slew Marcus Annius Verus to revenge James the Greater so his son Lucius Arrias Verus slew Phillip to revenge Marcus Annius Verus his father so Thomas slew Lucius Arrias Verus to revenge Phillip so Marcus Aurelius slew Thomas to revenge Emperor Lucius Arrias Verus his father. The remaining Apostle John Zebedee or John the evangelist did not join in the revenge of the priest. He was sentenced to life in prison and served in Pathos in Greece and died of old age.

Summaries

In this way the commander of the Apostles, became the keeper of the Herod Priest and emperor line that was killed by his Apostles, however was still the commander of the Apostles even though they were dead. The rest of Galilee would be the resurrection of the Apostles. In this way then the Lord would be god of the Priests and the Emperors, and so come to rule the Roman Emperor, because Nero was a suspect in the burning of Rome.

There are many ways to do this. Variance is found likely in the One Star General, Zacheus with or without innocence. In the Two Star Alphy revenges his sister, either by what he thinks is his father, or by his mother, but because they killed it is still law. Also the slake of Alphy by Mary wife of Cleopus to make the eldest son of Alphy and Mary, or if she had twins, or not or if Judas is older than James or not or if Simon is before or after the slake.

In the three Star there is variance about the order of sacrifice of the sons of Mary Alphaeus but in some texts Judas may kill only once, and then James and Simon could kill or not but likely James the lesser would be eldest to the slake to sacrifice the lord, revenged by his soldier brother Judas, and Simon the eldest of Cleopus also a sacrifice so of the sons of Mary Alphaeus would be the slaked lord the legitimate lord and finally the soldier lord and in the end Barsabbas becomes part of the loss in contest about the lot of the short straw while Mathias becomes the apostle, in the knight of Lots children. It would be wise to think that James the Lesser did not kill, and Cleopus revenged his son Simon, so James the Lesser would be the father of Mariam Arrius and so not have a killing to his credit. Also it makes sense the Alphy would not Kill Zachariah to revenge his wife's mother but that Herod Antipas would revenge his wife's rape by Zachariah.

There could be a way that involves Judas Iscariot but that puts it on the Priest. Likely what the lord did to take over the world was sealed by Judas Iscariot becoming a rabbi, the rapid heart beat when his mind discovers he has spilt his stomach out with a knife and there is no way for the temple of his body to survive.

The Survivors

The lord would have made this plan to save Mathew, because of the Gospel, and John Mark because of the Gospel, and John Zebedee because of the Gospel. In the Priest there is only one go up or grant of the blessing, so Jonas would be the sacrifice of the faith to make John Mark Peters Scribe.

Elizabeth, Cleopus, Mary Alphaeus, Perpetua, Bernice, Bernice, Mathew, John Mark. John the Evangelist

The Casualties

Zachariah, John the Baptist, Virgin Mary, Herodias, Alphaeus, Herod Antipas, James the Lesser, Herod Phillip, Judas Thadeus, Peter, Marcus Titus Flavius Sabinus, Andrew, Jonas, Gaius Sillius Calpurnius Domitius, Phillip. Arrius Antonius Calpernius Piso Barholomew, Arrius Antonius Calpernius Simon, Marcus Annius Verus, James the Greater.

House of David Side Casualties. 12

Levitical Priest Side Casualties, 9

Apostle Casualties   James the Lesser, Judas Thadeus, Peter, Andrew, Phillip, Bartholomew, Simon, James the greater, Thomas

Apostle Survivors   Mathew, John Zebedee

The Sins

First the Hebrew or House of David Side

Zachariah slept with another man's wife, who bore Herodias

John the Baptist had Herodias or the sin of incest

Alphaeus had sin of Matricide or Killing his mother and Adultery with Cleopus wife Mary

Anne Killed her daughter

Mary the Virgin Slaked

Judas Thadeus commited homicide to revenge his father

James the lesser commited homicide to revenge his brother

Simon committed adultery with another man's wife to make Miriam Arrias

Andrew Committed Homicide to revenge Simon

Jonas committed Homicide to revenge his eldest son

Jonus committed adultery to make John Mark the scribe of a gospel

Peter committed homicide to revenge his father

Bartholomew committed homicide of a wife to revenge Peter

James the Greater committed homicide to revenge Batholomew

Phillip committed homicide to revenge James the Greater

Thomas committed homice to revenge Phillip.

Second, the Levitical side of King Herod.

NOTE: The Levitical side did not actually destroy the criminal, but rather made a way to have each individual crucified at each post to control the world. They were all guilty of a mortal sin, except John the Evangelist who was given life in prison and Judas Iscariot who took his own life and is likely a Levitical Rabbi.

Herod Antipas committed homicide to revenge for a son in law committing adultery against his daughter

Herod Phillip committed homicide to revenge his brother

Herod Agrippa committed homicide to reveng his uncle

Marcus Titus Flavius Sabinus committed homicide to revenge the rape of his wife's mother

Gaius Sillius Calpurnius Domitius Piso commited homicide to revenge his father

Arrius Antonius Capernius Piso committed homicide to revenge his wife

Marcus Annius Verus committed homicide to revenge his father

Lucius Arrias Verus committed homicide to revenge his father

Marcus Aurelius committed homicide to revenge his father

The methodology of the derivation of the Take Over the World Spell

For each general or star, it must start with an us, Such as Zaccheus or Aphaeus, Thadeus, or Jonus, except the fifth star that is not part of us.

The blood shed must not kill the gospel books or the innocent.

The Apostles will take revenge on the Romans, who occupied the Holy Land, and justice must be maintained.

The order of the Roman people thru the making of each star, will be from King Herod and his line on down to Marcus Aurelius.

The faith is made, and the see of Rome is made, which opens to the see of Alexandria, or the idea of sharing Knowledge.

The sins are all atoned for, and the Christian Era is begun.

[26] Name the Stars.

(Generals)

Zacheus First Star, or Adjajjent (one us)

Alphaeus, who make Two Star, or Brigidier. (two us)

Thadeus, the three star or Major General (Three Us)

Jonus. The four star or General or (four us)

Not an us. And some Idiot put on five stars, which doesn't make any sense, so he can go to hell. So the Hellenism of Saint Helen makes cures out of the hell left by the commanding General.

[27]The Centurian Posts.

The Star of David is the one Star

The Centurian of Moses is the Two Star

The Centurian of Isaac is the Three Star

The Centurian of the Father Abraham is the Four Star

The Centurian of the Greek is Hercules or Helen is the Five Star

[28] During the War.

The Order of the Tribes is

Peter tribe of Simon Colonel and Naval Captain, heros Apollo and Ulysses

John tribe of Benjamin One Star hero Paris who kidnapped Helen

Bartholomew tribe of Dan Two Star and hero Perseus husband of Andromeda

James the Greater tribe of Judah Three Star and hero Persephone (wife of Hades)

Phillip tribe of Isaachar four Star, Hero Demetrius who married

Thomas tribe of Joseph Five Star Hero and God Dionysius

Order of Apostles during the War,

Order of Apostles

Mathew

Judas Thadeus

James the lesser

Simon

Andrew

Peter

Bartholomew

James the Greater

Phillip

Thomas

John the Evangelist

[29] War is Over

Takes place after Victory, and before peace is declared.

Five Star or Hercules is Retired or in disgrace, so Helen takes the Five Star to order the Cures made

Four Star is Reuben or Judas Iscariot

Three Star is Isaac-har or Phillip

Two Star is Levi or Mathew

One Star is Joseph or Thomas

Colonel is Benjamin or John the Evangelist

Lt. Colonel is Zebulon or James the Lesser

Major is Dan or Bartholomew

Captain is Asher or Simon

Lieutenant is Gad or Andrew

2nd Lieutenant is Naptoli or Judas Thadeus.

These are the ten tribes ordered by Ezekiel.

Both Judah and Simeon stay now in the Rear Admiral.

[30] Peace Time

Peace is declared and Victory is done, and the people return to life.

Judas Thadeus tribe of Naptoli Greek lady IO mother of Hero Epaphus rank 2nd Lieutenant

Mathew tribe of Levi Greek Lady Hera mother of Hero Hephaestus rank Lieutenant

Andrew tribe of Gad Greek Lady

James the Lesser tribe of Zebulon Greek Lady Deinara mother of Hero Hyllus rank Major

Simon tribe of Asher Greek Lady Ephrsus mother of Hero Minas rank Lt. Colonel

Peter tribe of Simon Greek Lady Leto mother of Hero God Apollo rank Colonel

John tribe of Benjamin Lady Hecuba mother of Hero Paris rank Adjujent Genera

Bartholomew tribe of Dan Greek Lady Danae mother of Hero Perseus rank Brigidier General

James the Greater tribe of Judah Greek Lady Demeter mother of Heroin Persephone rank Major General

Phillip tribe of Isaachar Greek Lady Rhea mother of Heroin Demeter rank General

Thomas tribe of Joseph Greek Lady Semele mother of Hero God Dionysius rank Commander General

Judas Iscariot tribe of Reuben Remaining Greek Ladies, and Heroes, and not a rank.

Apostle	Tribe	Greek Lady	Greek Hero	Rank
ANY	ANY	DRYOPE	PAN	LANCE CORPORAL
ANY	ANY	ALCEME	HERCULES	SARGEANT
JUDAS THADDEUS	NAPTALI	EPHRASUS	IO	2ND LIEUTENANT
MATHEW	LEVI	HERA	HEPHASUS	IST LIEUTENANT
ANDREW	GAD	MAIA	HERMES	CAPTAIN
JAMES THE LESSER	ZEBULON	DEINARIA	HYLLUS	MAJOR
SIMON	ASHER	LETO, EUROPA	MINOS	FIRST MATE, LIEUTENANT COLONEL
PETER	SIMEON	ANTICETUS, LETO	ULYSESS,APOLLO	CAPTAIN OF THE SHIP, COLONEL
JOHN	BENJAMIN	HECUBA	HECTOR	ADJUJANT GENERAL
BARTHOLOMEW	DAN	DANAE	PERSEUS	BRIGIDIER GENERAL
JAMES THE GREATER	JUDAH	DEMETER	PERSEPHONE	MAJOR GENERAL
PHILLIP	ISAAC-HAR	RHEA	DEMETER	GENERAL
THOMAS	JOSEPH	SEMELE	DIONYSIUS	COMMANDING GENERAL
JUDAS ISCARIOT	REUBEN	HARMONIA	ANY	HEBREW

[31] Apostles Veneration of Tribal Chieftains or the Spell of Making a Post

Apostle	Tribe Apostle Venerates	Veneration Number of Marriages	Marriages			Other Tribes
Judas Thaddeus	Naptali	2	Thalma wife of Neri	Anna wife of Abijah		All except Gad
Mathew	Levi	0	Mathew is the blood of Levi			All except Gad
Andrew of House Medes	Gad	Impossible				Unknown
James the Lesser	Zebulon	1	Barayah wife of Perez	Kanita wife of Hezron		All except Gad
Simon	Asher	Unknown 2 if Zibiah of Blood of Asher	Thalmar wife of Neri	Zibiah wife of Ahaziah of blood of Asher		All except Gad
Peter	Simon	Unknown				Unknown
John Zebedee	Benjamin	4	Salome; Joanne	Thalmar wife of Neri	Ahio of the line of Saul	All except Gad

Bartholomew	Dan	3	Thalmar wife of Neri	Ahio of the line of Saul	Hishim Granddaughter of Dan	All except Gad
James the Greater Zebedee	Judah	2	Salome married Judas Zebedee	JoAnne is male line of David		All except Gad
Phillip	Isaac-har	4	Miriam of line of King Herod;	Jehosebah daughter of Jerhoram; Athalia wife of Jefhoram	Ahijah wife of Nadab son of Jeraboam	All except Gad
Thomas	Joseph	2	Thalmar wife of Neri	Jecholiah wife of Amaziah or Athalia wife Queen of Jehoram		All except Gad
Judas Iscariot	Reuben	2	Jehosebah daughter of Jehoram	Rahab mother of Boaz of line of Reuben		Judah Zebulon Levi Isaac- har

Apostle	Literature	Country of Fate	Fate
Judas Thaddeus	Letter of Jude	Armenia	Martryed in Armenia

Mathew	Gospel of Mathew	Myrna Greece	Tradition is a Christian Literature the Martyrdom of Mathew In myrna by fire on a bed, with Bishop Plato
Andrew		Northwest of Black Sea	Crucified near Black Sea
James the Lesser	Letter of James	Israel	After Roman Govenor Festus Before Lucceaus Albinus A Great Fall From the Temple that injured and then stoned to death after a prophecy fullfillment in Jeremiah. Stoned by a Fuller
Simon		Israel	Brought before Roman Governor Atticus during Reign of Trajan and martyr in Holy Land
Peter	Gospel of Peter Letter of Peter	Rome Italy	Crucified in Nero's Garden
John	Gospel of John Gnostic book of John Revelations Letter of John 1 2 3	Greece Ephasus	Survived boiling oil, went to Patmos, Moved to Epheus, died at age of 94 in Epheus during Trajan
Bartholomew		Armenia	Flayed and beheaded by King Astyages in Armenia

James the Greater		Israel	Beheaded by King Herod Agrippa In Jerusalem
Phillip	Gospel of Phillip Apocalipse of Phillip	Greece	Crucified upside down during Emperor Domicletian in Greece
Thomas	Gospel of Thomas Gnostic book of Thomas	India	Crucified in India
Judas Iscariot	Gospel of Judas Iscariot	Israel	Suicide to make Rabbi in Potters Field
Heir to house of David in the Night Jesus Christ		Israel	Faked Crucifiction and lived to ripe old age

Strategic Positioning of Apostles during Disporia

It is known the the Lord of the Hebrews, Foster child of Saint Joseph in his second marriage or marriage of the night, was lord of most if not all of the kingdoms risen before the year one.

It is thought that the Lordship of the Christ was to unite all people in an effort to make a common government for the entire known world.

Consider the martyrdom of the apostles and lifes that they led after the crucifixtion of the lord and the sentencing of Nero in the Garden and how it was done.

[32] Idiot Spells

There are not parallel realities,   there are people with sophisticated computer programs that brain wash people with computers remote operation, and televisions, using subliminal suggestion and other intelligence tricks.   These people are dangerous idiots

Idi Amien ruler of Uganda, to discipline his prince son took his wife and cut off all limbs and switched left to right and right to left.

.

[33] Hand of War Spell

Father Bill said that a brother in the back of the truck kept his hand outside the truck.

Father driving almost hit the fence.

Brother hand was maimed, and he held it up and said I am truly blessed.

Reason for Kohath, son of Levi, the hand of war.

[34] Limp Dick Spell

Astarte Ningal Enkidu, NinMah, Anu, Anskar, Tiamet, Apophis, wife of Yam, Iluyanka

This spell is the difference about a Yank.   Possibly the origen of the Yank or Yankee term.   Made from the Sumatrian dieties. This spell will cause the victim to be completely hasseled and bothered with while a lady massages, sucks, or anything and the victim is continually distracted an cannot concentrate enough to get it up.

[35] Counter Limp Dick Spell

Astarte Ningal Enkidu, NinMah, Anu, Anskar, Tiamet, Apophis, wife of Yam, Iluyanka daughter Babys, Perecide son of daughter of Ion, Ion, Sol or Apollo.

To counter the Limp Dick Spell you must get away from Iluyanka. And go to the Greek God Apollo son of Zeus.

[36] Take a Dunk Man

Astarte Ningal Enkidu, NinMah, Anu, Enlil, Nergal, Enkidu, Ninmah, Adama

This spell is made out the sumatrian dieties, and will cause a man to deficate.

[37]Take a Dunk Women

Astarte Ningal Enkidu, NinMah, Anu, Enlil, Nergal, Enkidu, Nin-Khursag, Eve

Same as spell for Adamah, but is for Eve or Women.

[38] The Wind Spell

Astarte, Ningal, Enkidu, Ninmah, Anu God of the Sky, Enlil God of the Wind

This spell is a worship of the Jealous God Enlil. He is happy to make wind, sometimes in the strength of hurricanes or tempests.

[39] Spell of Division

When a deified God uses a hatchet to divide the black snake during the pregnancy of the mother subjected to more hel than any other women, who can assimilate the spirit of the serpent as well as the spirit of her noble father into the deified Gods Grandson, who is destined to make kingdoms topple and cure great diseases, stealing death from the powers that be and be more powerful than the possession the serpent uses, by possessing all gold and overcoming the pain of dispossession.

[40] Spells of Nursing

Rebecca Towne Nurse was an innocent lady, who was hung as a witch by Governor Phipps, in Salem Massachuesetts.   She was not a witch, she was not guilty of Killing Anyone.

Astarte, Ningal, Enkidu, Ninmah, Anu God of the Sky, Enlil God of the Wind, Ninlil the Beautiful Nurse

Rebecca had Twenty Four Cousins, including her cousin Deborah her nurse daughter of Huz.

[41] Jan Spell

Joseph as father of Janna the brother to his sister Janna who had Achim son of Zadok and Janna made Eliub who married Janna the daughter of Mathias and made Eleazor

[42] Spell of Doorway Lifting

Stand in a doorway.   Take both arms and press the back of your hands into the door Jam.   Use all your strength and hold for five minutes.

Step out of the Door, and relax your arms.   The spirit of the Doorway will lift your arms although not willed by the mind of the person who stood in the doorway.

[43] Spell to Defeat Asmoneus the Demon

Asomoneus was a levite that appartently turned evil.   He is mentioned in the book of Tobit, a sacred literature from the old testament were Azaziaz or The Archangel Raphael is mentioned as the person to defeat him.   Sarah had many husbands all killed by the demon Asamoneus.

Asamoneus is mentioned in the Anti Nicean father, a sect of Christianity that had arguments against what was declared true for the church by the council of Nicea,

Asmoneus father was the High Priest Onias the 47th High Priest.   Priests concerned in Asamoneus defeat is from Jaddua The 37th High Priest.

The priest must be completed in order from 37th, 38th, 39th etc.   It would be easy to do if the 38th priest was the father of the 39th and the 39th was father of the 40th etc,   this is not the case however.

Jaddua was the 37th high Priest.   The was succeeded by his son Onais the 38th high priest..   In the cool this is one the way. The 39th was the Simon the Just.   Now if Jaddua or Jade is the rock of royalty and his grandson is Sinon the just then everyone has to be treated justly. These are priests so that should go without says. The 40th high priest was the brother to Simon the Just, Eleazar. Eleazar was the 40th High Priest not difficult so far. Now 41 was the son of Jaddua Mannassah. Mannassah was succeeded by the son of Simon the Just. Now 42nd 43rd and 44th are direct sons of Simon the Just Onias the 2nd 42nd High Priest the father of Simon the 2nd the 43rd High Priest the father of Onias the 3rd the 44th High Priest. The 45th high priest was Jason the son of Simon the Just who is the 39th High Priest so you have to climb the lineage back to Simon the Just and down on generation to Jason High Priest. 46th High Priest was Menelaus the son of Simon the Just High Priest and brother to Jason. Now the 47th High Priest was Onias the 4th the father of Aaamoneus. So you see the problem in the succession of the High Priest.

In the Aaronic High Priest the succession was done by the most knowledgeable Priest to help the people. So it was not always from Father to son. The relationships in the succession of the High Priest would be very difficult to figure out. Thus would be difficult to follow.

Now Asamoneus is a direct ancestor of the Virgin Mary, and so also her sister Salome whose line continues to the King of Amorica Judiccal. So if a priest is angered all he has to do is put you in Asamoneus your ancestor and your Leviticus Is too difficult to figure out and basically you have to yield to the priest. Rafael the Arch Angel figured it out and now you can too.

Notice how many of the High Priest are easily shown to be our ancestors, however many like Jason the 45th High Priest are ancestors but not the same as most of the high priests because his son was not the High Priest at the time of his leaving the office, but a relative was.

Conclusions. In order to defeat Asamoneus you have to be able to climb 7 generations and father to children then climb back up to the father. Now the clencher.

The next High Priest, the 48th High Priest was not from the same line of Aaron like the house of the High Priest. He is Alcimus the 48th High Priest. The only requirement for a High Priest was to be descended from Aaron. The Levites usually stayed where they were all related. But Alcimus the next High Priest was an Aaronite however not from the lines like all the other high Priest.s In order to get to Alcimus then you have to go all the way back to Aaron and then down to Alcimus and to get to Jonathon Appus you have to go to Aaron Again and then to Mattathias the father of Johathon, Simon Thassi, and Judas Macceabeaus Elephant lord and great Warrior.

[44] Spells to Repel Demons for Mary Magdalene.

Now Mary Magdalene was possessed by seven demons according to the scriptures. These demons may be presented in the Gospel of Mary Magdalene that was found in the late 19th Century. These demons were darkness, Craving, ignorance, Lethal Jealousy, enslavement to the body, intoxicated wisdom, and guileful wisdom. There is a genealogy that explains the heresy and the truths and parallels that the heretical take advantage of.

The darkness curse is done with the iris of the eye. It can be judged to open or close to regulate the amount of light the subject sees. If closed enough the darkness can cause depression in the person due to the dim light.

Craving is difficult to counter, if about food possibly becoming a vegetarian, or if about sex, abstenance or satisfy the craving constantly until marriage or the lust is consumed.

The ignorance curse is the easiest to counter, however education or instruction takes a while.

Lethal Jealousy is difficult to contain, and if used will likely get someone killed. Not a good thing. Better to realize that if you really love someone that much then it is better to set them free because you love them.

Enslavement to the body is difficult, if its sex, ok, if its exercise, or if it's the idea you have to look good for your mate, then reason will likely repel this demon.

Intoxicated Wisdom is usually with alcohol or excessive work. To repel this demon either sober up or rest.

Guileful wisdom demon is repeled by humility, if you know you are not that smart then quit making arrangements for people.

[45] Spell to Resurrect the Dead

Dear Lord please resurrect the following person who has died (state the name) Post Christianity King Arthur who is blood of Heber Hyperion who marries Tamar the granddaughter of Josiah King of Judah who is a descendent of Jehosaphat a descendent of King David. Jehosaphat is married to Malika who is of the line of Gershom son of Levi. So from Malika navigate to Asaph who wrote many songs (straight up the line) who has many sons and is a boys choir, so the boys are the innocent and Michael the levite or Gershomite takes the dead soul from its carry and makes it part of the Pregnant Woman, whose child now has a soul that is both new and resurrected.

[46] Spell of Making the Lord

Heresy of Jesus Christ, the greatest Hebrew make
In order to make a king greater than David and Daniel then this king must be affiliated with Egyptians, Persians, Greeks, and the Roman Empire.

Now if Daniel is from two generations of Incest on both his mothers and fathers side and his parents were incestuous, then that much love would be hard to beat.   So the Roman Empire would be involved in creating a King of the Jews greater than King David, the father of Daniel, and yet also a Priest.

Now everyone knows that the Virgin Mary is the Mother of Jesus Christ.   Her only brother was Alphaeus.   Now we know that Ann was the mother of Mary, and also the mother of Alpheaus. Ye David had two generations of Incest not one.   So Anne would have to marry her half brother to make the Virgin Mary and Alpheaus.   The Virgin would be born first and then Alphaeus. Joachim would have Ann and make Salome, since his prayer ended up with Mary and the Temple.   So Anne's brother was Bar-Jesus the son of Bianca and Illegitimate, half brother to Anne who was also from Bianca.   Now Anne father was likely Joidiah so that Anne would be from the line of Aaron.   Bar Jesus father was the son of Bianca by Jeshua III, however has been called illegitimate from some sources. Now consider that Joseph of Arimathea was the brother to Bianca. So he is likely the father of Bar Jesus, not Jeshuah.   That's three generations of Incest that is better than Two, like Daniel. So the Hebrew have a greater make in Jesus Christ rather than King David.   King David is still the King over the jews, unless they go into Christianity.

The graphic shown below is likely part of a Heresy. It is certainly heretical and uses a different way of thought than what conventional Christians use. It would have to be called a heresy if someone who is not ready for this encounters it. However it is likely the truth.

So the make of a Lord is going to take a few generations. This is Similar to the make of David that also involves incest.

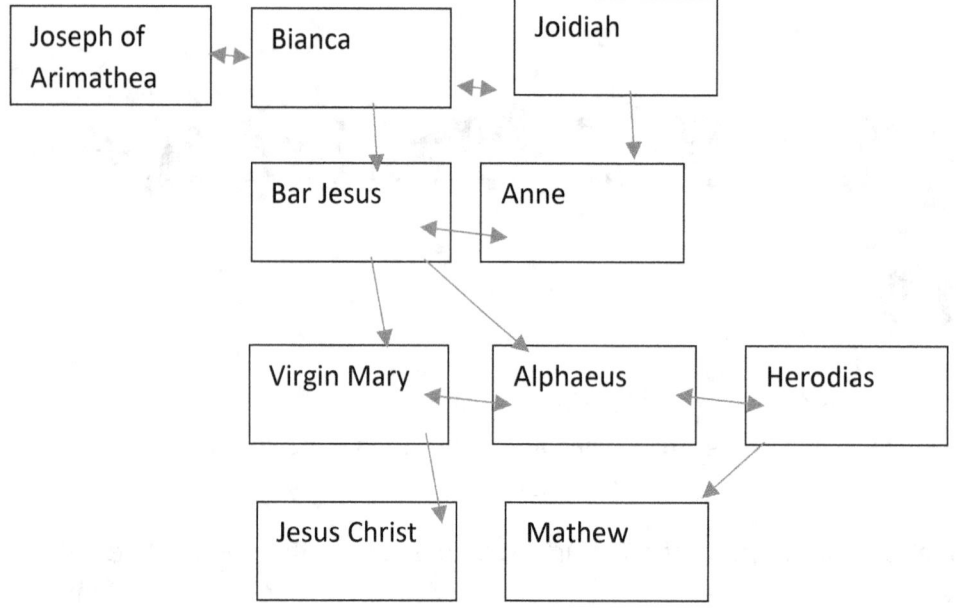

[47] Spell of Love Arimathea

The spell of Arimathea requires a Post of Saint Joseph of Arimathea and a correct veneration from Saint Caradoc of the Lord. These lines are hidden so that everyone who calls on the name of the Lord does not start a Arimathia.   Also Medical doctors may prescribe heart medicine for an irregular heartbeat. Before Casting this it is a good Ideat to Post Saint Anne, and also make sure the lover is healthy enough. Also not for Children . Most of this is for the Mother and Father of whoever so that the Arimathea will generate enough love to last thru the night.

Jesus Christ, In the Resurrection

The graphic shown below is likely part of a Heresy. It is certainly heretical and uses a different way of thought than what conventional Christians use.   Here the Enygenus marries the Lord and has a child Carodoc. For the Neophyte Carodoc is son of Bran (Bran was Duped) and Ann of Arimathea. The Resurrections of the Lord follow (Shown in Bold) so it eventually merges with the male line of Judah to form Strathcylde.   It has to be presented as a heresy because the old and infirm and the children could be hurt if an arithmea is started.

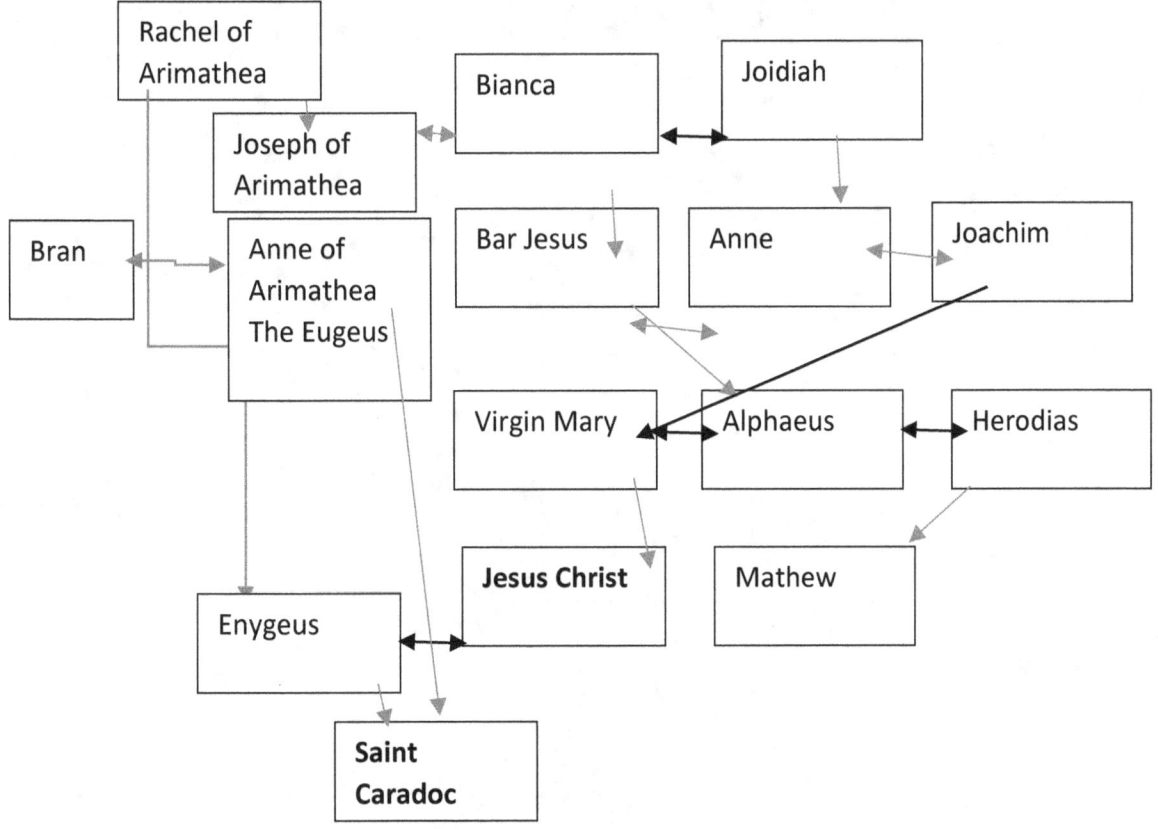

Navigate to Saint Anne of Arimathea.   Check the Euygeus and go to Rachel of Arimathea. If She wants you to feel love of Arimathea, she will post Anne the Priestess daughter of Joidiah the Levite and then post the true Eugeus from Caradoc and the line of Jesus the crucified son of Alphaeus son of Bar Jesus Son of Saint Joseph of Arimathea.

[48] Spell to Post the Greatest Christian King

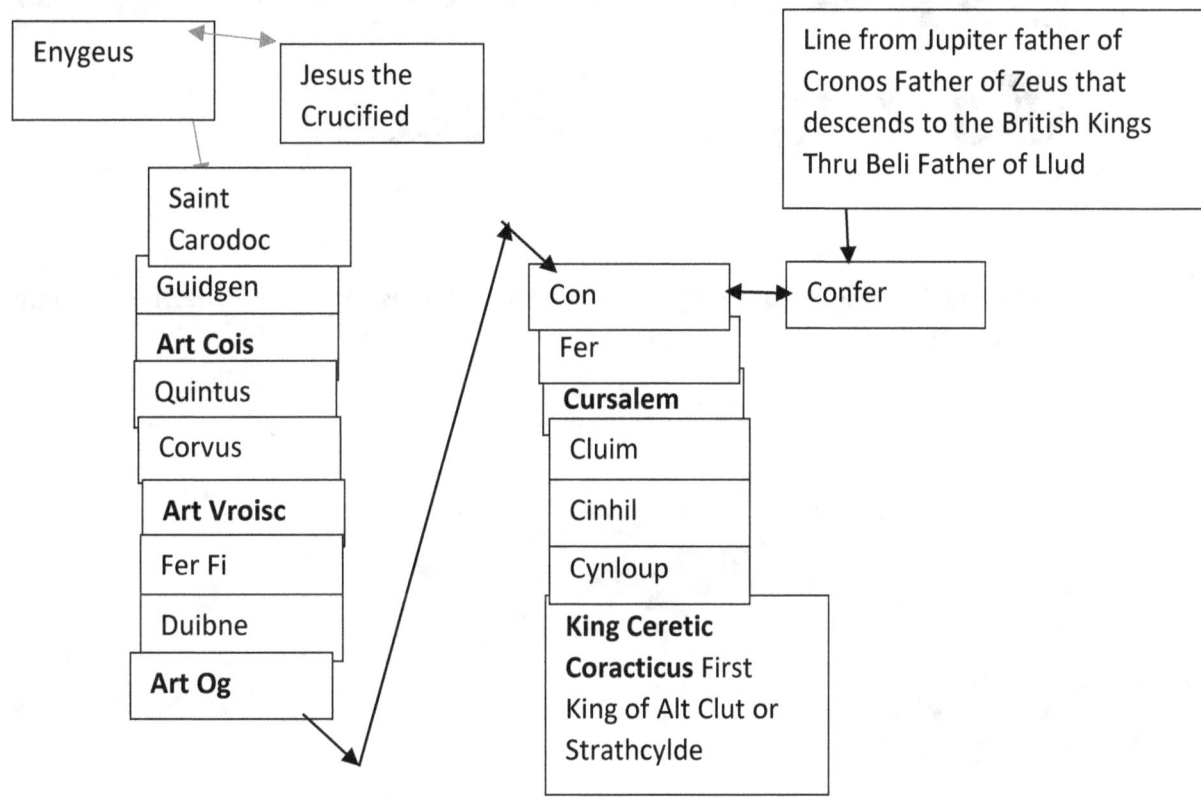

Saint Carodoc Grandson in the Resurrection Art Cois in the Resurrection Great Grandson Art Og in the resurrection CurSalem Add the mighty Arm of your Grandmother Line of Zeus to your High Priest Hilkiah. Resurrection of CurSalem and the Lord with the mighty Arm of Zeus pull your Hilt and Cleve to fulfill the Law and Take Dumbarton Rock, King Ceretic Coracticus Rule your Davidic Kingdom with Justice and Law and give to all invaders their Tic of Slavery.

[49] Spell of Confusion

There is no invisible spell here.

[50] Spell To Throw Enemy To Chaos [Greek]

Great God Zeus rise to your Father Cronos son of Uranus son of Gaia The Titanus to the Great Titan Chaos.

[51] Spell to Throw Enemy to Chaos [Israelie]

Judah son of Israel the way and the light rise to Israel and his wife the Daughter of the Sorceror and Israel rise to your Great Mother who is the faithful Intelligence and to the Child of the Promise Chaos (Isaac)

[52] Spell to Pass Thru Chaos and Create Voluptous Women

Great God Zeus rise to your Father Cronos son of Uranus son of Gaia The Titanus to the Great Titan Chaos Father of Eros Father of Erato who has Pschye Parents of Voluptua and the thought makes a Voluptus Women after many years.

[53] Spell to Throw Enemy to Chaos [Sumatrian]

Astarte Great Goddess your husband Cush Son of Ham Go to your Sacred Mother Emzara wife of Chaos (Noah)

[54] Spells To Rescue Oneself from Chaos

(Greek and Hebrew)   Chaos, Domnu , Descend to Dalbreach Father of Elathan Father of Ogma Father of Deleath and Bridgid mother of Donann who with Deleath made Brian or Orihah father of Kip whose line descends to Omar whose line descends to Heth Father of Uriah whose daughter marries the Dark Angel Uriel Angel of death whose daughter marries Rehoboam Grandson of King David.

[Sumatrian] Chaos, Father of Domnu marries The Granddaughter Of Olivia And Ham Father of Cush who has Astarte who has Dumuzi whose line Fathers Gordia who marries Cybele.   Great Goddess Astarte befuddle Cybele from her diabolical design and let the descent to Lydia continue to King Ardys of Lydia rise to Pluto daughter of Himas who marries the Daughter of Hades son of Cronos.

[55] Spell to make the Great King

Great Wizard Ibzan, son of Ruth and Salmon father of Edal Father of Abala cast to Jesse who posts his Gabe thru the Nether regions of the Galaxy to the Red Rays of Love that lead to the center of the make the Great King of Rigel and of the Milky Way David.

[56] War Spell

Mary Magdalene Mathias Simon Thassi, to post Miriam Maccabeus, to Mathias, to Judas Macabeus, move Ollie (First Macabees),   move Macy (Second Maccabeus),   Dumbo Awake (fourth Maccabeus)   keep the rhythm and the ryme and row (Third Maccabeus) and route the herd until they trample the enemy, row the lines, and when he slips, call Obidiah to Zerubabel, carry to Shazzrazah, to Nehemiah for the walls of Jerusalem to be rebuilt.

[57] Spell of Blessing

Judah, Bathshua, Hasadiah, Zedekiah, Masseiah, Neriah, Baruch (Baruch in this language at this time means Blessed so his name is Blessed)

[58] Spell to Raise the Sun

Astarte, Ningal, Enkidu, Ninmah, Anu, Anskar, Tiamet, Apophis, daughter of Yam, Iluyanka, daughter, Baby, Perecides, son of Ion, Ion, Sol, or Apollo

[59] Spell to Raise the Devil

Naptali, to Suri who has Sarvia the harlot and makes Joab whose daughter has Bichri father of Sheba.

Reuben to Hezron to Caleb to Hur and Miriam to Ezer and Miriam to Chelob, to Mehir to Tehenneh to Nahash and Abala to Sarvia.

Joseph the Israelite, Epriaim, descend to Azaziaz who has Sheba from the big Sur and makes Nebat who has the Harlot Sarvia, who makes the Devil himself.

[60] Spell to Raise Death

Jacobs son Levi, Kohath, Amram, Jocabed and Elizabeth mother of Nadab. (Death) with his brother Abihu.

[61] Spell to Raise Judgement

Ket Serug, Nahor, Terah, Abraham, Isaac, Israel, Dan or the divine Judgement

[62] Spell to Raise the Magician

Ket, Serug, Nahor, Terah, Abraham, Isaac, Israel, Joe the Magician and his brother Ben.

[63] Spell to Hang Ten Man

Ket, Serug, Nahor, Terah, Abraham, Isaac, Esau and Basemath, Timna, Amalek, Apag,Hammedatha,Haman father of the hanged, Dalphon, Poratha, Aspatha, Aridatha, Arisai, Adalia, Vajeezatha, Parshandatha, Parmashta, Aridah

[64]   Spell to Raise the Staff

Ket, Serug, Nahor, Terah, Abraham, Isaac, Israel, Judah, Perez, Hezron, Jathir (Wood)

[65] Spell to Heighten For Fisticuffs.

King David, Abigail, Jeter, Jada, Onam, Jeremeel, Hezron (Right Cross)

[66] Spell to Make the Martialis win the contest

Jesse, Abigail, Daniel, to Hilkiah, to Gamaliel post Simon father of Imme Shalom, and From Mathan, Hizkiah, Judas the Zealot, Simon, Hyrancus marry the Imme Shalom, descend Eleazar Ha Gadol, Simon, Gamaliel, Hillel, Malchion, son, Conaus the Martyr, son Salomus (Shallom, Salomon) Urbanus, Mansuetus, Symetrius, Martialis,------- Strike

The second blow is an insult to the Saint, because the first Blow is Fatal

[67] Spell In the Knight to Add Food and Shelter to the Pregnant Women with Child

Mathan, Jacob, Eucheria, Miriam, Theudas, Addai James, Unknown Father of Sotor,

Unknown Selius goes to Ptolas or Bartelmaeus, and buys the beasts in Worship of the Man who proved he was god of that beast, and then to Ham and Olivia to Domnu to Add the Songs of Life to the child and then to Paratama the Overlord, and to Shaushatar the wood Lord and to Aratama the Horse Lord, and to Shuttarna the War Lord and to Tushratta the Axe Lord.

[68] Spell to Cure Arthritus

Saint, Saint, Saint, Joe of Arthimea in the right to Saint Joseph of Ariamathea

[69] Spell to Open
the ear when the skull closes on it

When the sin of the father is so bad those in judgement cause the bones of the skull to close the ear canal,   then this spell will open it and allow the Priest or Rabbi to see the sins of the subjects father.

Ket, Serug, Nahor, Terah, Abraham, Isaac, Israel, Rueben, Hezron, Caleb, Hur, Epratha

Ket, Serug, Nahor, Terah, Abraham, Isaac, Israel, Joseph, Mannassah, Machir, Epratha.

[70] Spell to Cure Cancer

Oh Great King and Saint Ethelbert Please cure this Cancer.

Saint Ethelbert believe in God Odin and ask the favor of God's Wife Frigg whose lineage marries Buonduah whose lineage marries Ann the Daughter of Jose the Rama Thea the son of the Saint Joseph of the Davidic Kingdom.   Saint Joseph post your daughter Ann and ask your grandfather Mathan favor with his daughter Don who marries Mathan Great Grandson Beli.

[71] Spell to Raise the Tempest

Charlemagne   Great emperor *Karolus serenissimus augustus a Deo coronatus magnus, pacificus imperator, Romanum gubernans imperium, qui et per misericordiam Die rex Francorum atque Langobardorum* let your line descend to the Vermandois to the Brittany to King William Conquer !!!!!   King Henry of England, Meshenines, Tempest Rise and consume the enemy.

[72] Take Over the World Spell thru the Son

House of David, post the formation of Christianity and the feat of Don marries Beli, 4 Generations

Post the Big Sur the feat of Sarvia mothers Jeroboam, 5 generations

Post the Healing Angel Rafael whose sister Sarah Marries Tobias 6 Generartions

Finally increase the Son Worship where Yakobom Grandson of Pepi marries the ninth generation daughter of Rahotep   7 Generations.

Long Live Ra the Son.

[73] Spell to put a Rainbow in the Sky

O great Judah!!!!   Bathshua, Shelah, Nafut, Sonas, Abilit marries Obed

Or

O Great David, Jesse, Obed marries Abilit, Sonas, Nafut, Shelah, Bathshua, Judah

[74] Death Spell in the Knight

Saint Joseph, let your father revenge your enemy son of Jacob who marries Eucheria mother of Miriam who marries Theudas, son of Anthrongonies the Good Shepard father of Amethas Perisha

[75] Naked Kill Spell in the Knight

Saint Joseph, let your father revenge your enemy son of Jacob who marries Eucheria mother of Miriam who marries Theudas, son of Anthrongonies the Good Shepard father of Hanibas

[76] Raise the Priest Spell in the Knight

Saint Joseph, let your father revenge your enemy son of Jacob who marries Eucheria mother of Miriam who marries Theudas, son of Anthrongonies the Good Shepard father of Amran

[77] Raise the Heroes in the Knight

Saint Joseph, let your father revenge your enemy son of Jacob who marries Eucheria mother of Miriam who marries Theudas father of James, father of Evodius Father of Heroes One father of Cornelius, father of Heroes Three.

[78] Spell of Levitical Love from King Herod

King Herod married Marriame mother of Alexander of Judea and King Herod was father of Archelaus who married Marriame I mother of Glaphyre of Cappadocia who married Alexander of Judea father of Tigrantes IV who married Erato of Armenia.

[79] Spell of Casting Lighting

O Great Jupiter, see the victim now named AEsculapius, and now named Tullus Hostilius,and now named Homogyrus, all victims of your great arm of lightning, cast now on the victim and send him to Hades as Zorastor throws the bolt.

rnb

The following spells are names of God given to us by the father as triads.

[80] Spell to Remove Negative Thoughts

O Great Father Abraham, remember thy name of God r n b

[81] Spell to Arouse Total Certainty

O Great Father Abraham remember thy name of God rvk

[82] Spell to Arouse Healing Powers

O Great Father Abraham remember thy name of God o e x

[83] Spell to remove Negative Forces

O Great Father Abraham remember thy name of God L E Z

[84] Spell to Generate the Energy of Financial Sustenance

O Great Father Abraham remember thy name of God Q A N

[85] Spell to Remove Egomania

O Great Father Abraham remember thy name of God A P K

[86] Spell to Eradicate Death

O Great Father Abraham remember thy name of God P K Z

[87] Spell to Return to Seed Level of Existence

O Great Father Abraham remember thy name of God F E F

[88] Spell to Stand After we Fall

O Great Father Abraham remember thy name of God E U B

[89] Spell of Courage to Speak and Hear the Truth

O Great Father Abraham remember thy name of God P Z E

[90] Spell of He May Add

O Great Father Abraham remember thy name of God A N D

[91] Spell of Oneness and Completeness

O Great Father Abraham remember thy name of God O N E

[92] Spell of Strength

O Great Father Abraham remember thy name of God O K E

[93] Spell of the Earth Spirit

O Great Father Abraham remember thy name of God E A A

[94] Spell of Robbery

O Great Father Abraham remember thy name of God R O B

[95] Spell to Put Lucifer in someones Mind.

King Hugh Lusignan the Fourth was Father of Hugh Lusignan the Fifth was Father of Hugh Lusignan the Fourth was Father of Hugh Lusignan the Fifth was Father of Hugh Lusignan the Sixth.

(It is impossible to have a son be father of his father, but if this is not caught it puts error in the mind.)

[96] Spell of the Lucky Seven

Post King David Father of Solomon Father of Rehoboam Father of Sharmariah.   This is the Luck, Navigate to Sharmariah (see other luck spell).   Then post Tabael wife of Sharmariah, and note Tabael is Eight Generations up from Rehoboam while Sharmariah is one generations. Subtract and make the Lucky Seven.   Note that Sharmariah's line that continues has also the Descendents of King David from the line of the Kings of Judah.

[97] Spell to Make the Saducee Priest

The Pharisee Priest is based on the number 7. Use the spell above, the Spell of the Lucky Seven to apply the seven to the generations of Abraham.   Remember the recorded text that is sacred says there are fourteen generations from Abraham to David, Fourteen from David to the Babylonian Exile, and Fourteen to the Galilee.

So (1st) Abraham, (2nd Seven) Aminidab, (3rd Seven) David, (4th Seven) Ahiziah, (5th Seven) Manasseh, (6th Seven) Rhesa, or Meshallum (Tobits line), or Hananiah, Obadiah's Line Esli. Then (7th Seven) Levi, then past the fall of the Temple.

This is the Primary Priest the Pharisee use to monitor the Father using Solomon's Line, the Kingdom Line. Also note to get from Manasseh to Rhesa then an adopt has to be used because Solomon Blood line did not continue.

Note the 7th Seven is Levi. (A Priest)

[98] Spell to Make the Pharisee Priests.

The Pharisee Priest is based on the number 7. Use the spell above, the Spell of the Lucky Seven to apply the seven to the generations of Abraham.   Remember the recorded text that is sacred says there are fourteen generations from Abraham to David, Fourteen from David to the Babylonian Exile, and Fourteen to the Galilee. Since the line of Solomon ran out that is the Pharisee and only Pharisee Priest. But the Sadducee has the line of Nathan, which has many branches, but only on has the Christ, however all have Nathan in the main bloodline.

Priest of the Christ

So (1st) Abraham, (2nd Seven) Aminidab, (3rd Seven) David, (4th Seven) Joseph, (5th Seven) Her, (6th Seven) Zerubabel, (7th Seven) Maath, (8th Seven) Janna, (9th Seven) Joshua or (Jesus the Christ   continues to Alt Clut

Priest of Obadiah

Also So (1st) Abraham, (2nd Seven) Aminidab, (3rd Seven) David, (4th Seven) Joseph, (5th Seven), Joshua (6th Seven) Shealtial, (7th Seven) Hananiah, (8th Seven) Sheconiah, (9th Seven) Hunna, (10th Seven) Kahanna, (11th Seven) Hananiah or Hisdai   (both sons of Bostani) continues to Wessex

Priest of Tobit

Also So (1st) Abraham, (2nd Seven) Aminidab, (3rd Seven) David, (4th Seven) Joseph, (5th Seven), Joshua (6th Seven) Shealtial, (7th Seven) Elias, (8th Seven) Jeshua, (9th Seven) Sanatrouke, (10th Seven) Chosroes   continues to Pavlav in Armenia

Priest of James

Also So (1st) Abraham, (2nd Seven) Aminidab, (3rd Seven) David, (4th Seven) Joseph, (5th Seven), Joshua (6th Seven) Shealtial, (7th Seven), Eliub, (8th Seven) Elzarus (9th Seven) Ysayes, (10th Seven) Aldroneus, (11th Seven)Alain II (12th Seven) Berenger   Contains the Judicael

Priest of Sirach

Also So (1st) Abraham, (2nd Seven) Aminidab, (3rd Seven) David, (4th Seven) Joseph, (5th Seven), Joshua (6th Seven) Shealtial, (7th Seven), Sirach, (8th Seven) Honeseh

Note that this puts the Saducee Priest able to put any Lady into Joshua, who is likely the greatest lover jew

[99] Spell to Make the Saducee Priest with Christianity

The Pharisee Priest is based on the number 7. Use the spell above, the Spell of the Lucky Seven to apply the seven to the generations of Abraham. Remember the recorded text that is sacred says there are fourteen generations from Abraham to David, Fourteen from David to the Babylonian Exile, and Fourteen to the Galilee.

So (1st) Abraham, (2nd Seven) Aminidab, (3rd Seven) David, (4th Seven) Joseph, (5th Seven) Her, (6th Seven) Zerubabel, (7th Seven) Maath, (8th Seven) Janna, (9th Seven) Joshua or (Jesus the Christ), (10th Seven) Fer Fi, (11th Seven) Cinhil, (12th Seven) Beli, (13th Seven) Rhydderch, (14th Seven) Mael Coluim I father of Owain the Bald

This is the first line that has 14 sevens.

So (1st) Abraham, (2nd Seven) Aminidab, (3rd Seven) David, (4th Seven) Joseph, (5th Seven) Her, (6th Seven) Zerubabel, (7th Seven) Maath, (8th Seven) Janna, (9th Seven) Joshua or (Jesus the Christ), (10th Seven) Fer Fi, (11th Seven) Cinhil, (12th Seven) Beli, (13th Seven) Dyfal.

The fourteenth seven line would be Malcolm II (Malcoim X) son if he was ever made. It is not known that he had a son, he probably did not. But He did have a daughter Bethoc of Scotland that would be the 14th Seven. Because there are really only thirteen this would be the Bakers Dozen.

[100] Spell to Summon a Lover

Rahab, Bathshua, Aribath, Sarvia, use your sense of my worth to know you can be hired, and check that there is enough for me after your hire, so we may love with strength. Find me where I am.

[101]2nd Spell to Make the Pharisee Priest with Christianity

The Pharisee Priest is based on the number 7. Use the spell above, the Spell of the Lucky Seven to apply the seven to the generations of Abraham. Remember the recorded text that is sacred says there are fourteen generations from Abraham to David, Fourteen from David to the Babylonian Exile, and Fourteen to the Galilee.

So (1st) Abraham, (2nd Seven) Aminidab, (3rd Seven) David, (4th Seven) Joseph, (5th Seven) Her, (6th Seven) Zerubabel, (7th Seven) Maath, (8th Seven) Janna, (9th Seven) Joshua or (Jesus the Christ), (10th Seven) Fer Fi, (11th Seven) Cinhil, (12th Seven) Beli, (13th Seven) Rhydderch, (14th Seven) Mael Coluim I father of Owain the Bald

This is the first line that has 14 sevens.

So (1st) Abraham, (2nd Seven) Aminidab, (3rd Seven) David, (4th Seven) Joseph, (5th Seven) Her, (6th Seven) Zerubabel, (7th Seven) Maath, (8th Seven) Janna, (9th Seven) Joshua or (Jesus the Christ), (10th Seven) Fer Fi, (11th Seven) Cinhil, (12th Seven) Beli, (13th Seven) Riderch (14th Seven) Kenneth II MacAlpin

So (1ˢᵗ) Abraham, (2ⁿᵈ Seven) Aminidab, (3ʳᵈ Seven) David, (4ᵗʰ Seven) Joseph, (5ᵗʰ Seven) Her, (6ᵗʰ Seven) Zerubabel, (7ᵗʰ Seven) Maath, (8ᵗʰ Seven) Janna, (9ᵗʰ Seven) Joshua or (Jesus the Christ), (10ᵗʰ Seven) Fer Fi, (11ᵗʰ Seven) Cormac (12ᵗʰ Seven) Llywarch or (12ᵗʰ Seven) Elisedd

The fourteenth seven line would be Malcolm II (Malcoim X) son if he was ever made. It is not known that he had a son, he probably did not. But He did have a daughter Bethoc of Scotland that would be the 14ᵗʰ Seven.

[102] Spells of the Resurrection of the Father

Each Spirit would be a resurrection in the Great house. Thus the Great Great Grandfather would be the resurrection

and four generations different.   The Priest is separated by eight generations. So the Resurrection does not follow the Priest

Here they are.

.

So (1st) Abraham, (1st) Judah(2nd Seven) Aram,(2nd}Boaz (3rd Seven) David,(3rd)Abijah (4th Seven) Jehoram, (4th)Amaziah,(5th Seven)Ahaz,(5th)Amon (6th Seven) Amon, (6th)Rhesa, Then (7th Seven) Levi, then past the fall of the Temple.

Note from Ahaz the 6th Seven there are alternate paths. (7th Seven) Meshallum (Tobits line), or (7th Seven) Hananiah, (7th)Obadiah, Esli. Also lines to (7th Seven) Ireland Faidh, (7th Seven) Ugain Mor, Also (7th Seven) Irial King of Ireland, (7th Seven) Heber Hyperian of the line of Heber Scot

This is the Primary Priest the Pharisee use to monitor the Father using Solomon's Line, the Kingdom Line. Also note to get from Manasseh to Rhesa then an adopt has to be used because Solomon Blood line did not continue.

Note the 7th Seven is Levi. (A Priest)

Note that liberties can be taken, using the Priest and then the resurrection.

In common to all these lines is the Resurrection to David and then to Neri.

Solomon's Line

Abraham, Judah, Aram, Salmon, Jesse, Rehoboam, Jehosaphat,Joash,Jotham, Mannasseh, Jehoikim, Pediah

Nathan's Line

Abraham, Judah, Aram, Salmon, Jesse, Mattatha,Eliakim, Judah, Mathat, Joshua, Cosam, Neri, Abiud, Isreal, Eliud, Joseph

Please note on the following Lines that the Resurrections are different from the Priests, but the Priests are based on eight generations, while the resurections are based on seven. Using some statistics every four priests is then a resurrection.

Resurections of the Christ Line

So (1st) Abraham, (1st)Judah (2nd Seven) Aram, (2nd)Boaz (3rd Seven) David,(3rd) Menna (4th Seven) Joseph, (4th)Levi (5th Seven) Her, (5th) Addi (6th Seven) Zerubabel, (6th) Yoda (7th Seven) Maath, (7th) Nahum(8th Seven) Janna,(8th)Levi (9th Seven) Joshua or (Jesus the Christ continues to Alt Clut

Resurections of Obadiah Line

So (1st) Abraham, (1st)Judah (2nd Seven) Aram, (2nd)Boaz (3rd Seven) David,(3rd) Menna (4th Seven) Joseph, (4th)Levi (5th Seven), Joshua (5th)Cosan (6th Seven) , Neri, (6th) Hananiah (6th Seven) Hizkiah, (7th) Arnan (8th Seven) Shemaiah, (8th) Akkub (9th Seven) Hannan, (9th)Nathan (10th Seven) Zutra,(10th) Mar Zutra II (11th Seven) Bostani (11th)   continues to Wessex

Resurections of Tobit

So (1st) Abraham, (1st)Judah (2nd Seven) Aram, (2nd)Boaz (3rd Seven) David,(3rd) Menna (4th Seven) Joseph, (4th)Levi (5th Seven), Joshua (5th)Cosan (6th Seven) , Neri, (6th) Hashubah, (7th Seven) Tobit, (7th)Antigone (8th Seven) Jose (8th) Panthera line continues to Joanne and Zebedee

So (1st) Abraham, (1st)Judah (2nd Seven) Aram, (2nd)Boaz (3rd Seven) David,(3rd) Menna (4th Seven) Joseph, (4th)Levi (5th Seven), Joshua (5th)Cosan (6th Seven) , Neri, (6th) Hashubah, (7th Seven) Tobit, (7th)Antigone (8th Seven) John (8th) Eudamas (9th Seven) Tausorpis (9th) Sanatroupke (10th) Seven Khursraw I   line continues to Pavlav

Resurections of James

So (1st) Abraham, (1st)Judah (2nd Seven) Aram, (2nd)Boaz (3rd Seven) David,(3rd) Menna (4th Seven) Joseph, (4th)Levi (5th Seven), Joshua (5th)Cosan (6th Seven) , Neri, (6th) Abiud , (7th Seven), Eliub, (8th Seven) Isreal (8th) Eliub (9th Seven) Joseph (9th) Elzarus, Narpus, Helyas, Gerentonus, Brendan or Brother Gradlon Mar, Budig, Alain, Alain II, Daniel II, or Budic II, Nominoe

Resurections of Sirach

So (1st) Abraham, (1st)Judah (2nd Seven) Aram, (2nd)Boaz (3rd Seven) David,(3rd) Menna (4th Seven) Joseph, (4th)Levi (5th Seven), Joshua (5th)Cosan (6th Seven) , Neri, (6th) Hashubah, (7th Seven) Haggai, (7th) Eleazar   (8th Seven) Levi (8th) Honeseh Continues to Mary Alphaeus or Zebedee

Note that this puts the Saducee Priest able to put any Lady into Joshua, who is likely the greatest lover jew

Spell to Make the Pharisee Priest with Christianity

The Pharisee Priest is based on the number 7. Use the spell above, the Spell of the Lucky Seven to apply the seven to the generations of Abraham.   Remember the recorded text that is sacred says there are fourteen generations from Abraham to David, Fourteen from David to the Babylonian Exile, and Fourteen to the Galilee.

Line of King Dumugal

So (1st) Abraham, (1st)Judah (2nd Seven) Aram, (2nd)Boaz (3rd Seven) David,(3rd) Menna (4th Seven) Joseph, (4th)Levi (5th Seven) Her, (5th) Addi (6th Seven) Zerubabel, (6th) Yoda (7th Seven) Maath, (7th) Nahum(8th Seven) Janna,(8th)Levi (9th Seven) Joshua or (Jesus the Christ Art Cois,(10th Seven) Art Vroisc, Art Og,Cursalem,Cynloup, Dumugual, Beli,Beli, Owain. Artgal, Owain, Owain the Bald

Second Line of King Dumugal

So (1st) Abraham, (1st)Judah (2nd Seven) Aram, (2nd)Boaz (3rd Seven) David,(3rd) Menna (4th Seven) Joseph, (4th)Levi (5th Seven) Her, (5th) Addi (6th Seven) Zerubabel, (6th) Yoda (7th Seven) Maath, (7th) Nahum(8th Seven) Janna,(8th)Levi (9th Seven) Joshua or (Jesus the Christ

Art Cois,(10th Seven) Art Vroisc, Art Og,Cursalem,Cynloup, Dumugual, Beli,Beli, Eugain, Rhun, Dyfnwal II, Malcolm II Mac Kenneth

Line of King Erp

So (1st) Abraham, (1st)Judah (2nd   Seven) Aram, (2nd)Boaz (3rd Seven) David,(3rd) Menna (4th Seven) Joseph, (4th)Levi (5th Seven) Her, (5th) Addi (6th Seven) Zerubabel, (6th) Yoda (7th Seven) Maath, (7th) Nahum(8th Seven) Janna,(8th)Levi (9th Seven) Joshua or (Jesus the Christ Art Cois,(10th Seven) Art Vroisc, Art Og,Cursalem,Cynloup, Geraint, Bledwig, Cwflyn,Dfauwal, Canordoly, Hopkin, Elynd,Vortegyn,Juhel

Line of King Cinuit

So (1st) Abraham, (1st)Judah (2nd   Seven) Aram, (2nd)Boaz (3rd Seven) David,(3rd) Menna (4th Seven) Joseph, (4th)Levi (5th Seven) Her, (5th) Addi (6th Seven) Zerubabel, (6th) Yoda (7th Seven) Maath, (7th) Nahum(8th Seven) Janna,(8th)Levi (9th Seven) Joshua or (Jesus the Christ Art Cois,(10th Seven) Art Vroisc, Art Og,Cursalem, Cynloup, Tutugual, Neithon, Anllech, Anarawd Gwalch

[103] Spell to Ignite a Star

Simple Algebra.

Note the equation to calculate the force of repulsion between a Proton and a Neutron.

F=k*Q1*Q2/r^2

Force is equal to a Constant k times the charge on particle 1 and the charge on particle 2 divided by the distance between squared.

F= G*M1*M2/r^2

F is force between masses.

G is the Gravitational Constant

M1 is the First Mass

M2 is the Second Mass

R is the distance between the centers of the Mass

Now the Trick   F=F

Force from the first equation is the same as Force from the Second Equation

Therefore the force of gravity between the particles will overcome the electron force to repel the proton.

K*Q1*Q2/r^2=G*M1*M2/r^2

or

M1*M2=k*Q1*Q2/G

Solve for M1*M2 or Total Mass   This is Roughly the Mass of Jupiter the Morning Star

[104] The Spell to Arouse a Man

Great Goddess, let your husband Amenhemet father of Sensuret II who marries Nefru father of Meribah let the water flow, while her mother marries Tjenna mother of Tetisheri who marries Semkearee Tao I the wolf. (the song of the wolf is very much like the song of man).

[105] The Take Over the World Spell

The Take Over the World Spell

First Star John the Baptist

Zacheus or Zachariah married Elizabeth and made John the Baptist. Zacheus seduced and had Bernice, who was married to Aristobulus, so Herodias was born from Zacheus and Bernice and John the Baptist who liked her so together they fathered Mathew however John the Baptist was innocent since he did not know she was his sister. So Aristobulus slew Zacheus in the doorway for adultery.

Second Star Alpheus and Mary Alphaeus

So Mathew killed Aristobulus for killing his Grandfather and Salome found out and she had

Herod Antipas who was married to Herodias and he found out John the Baptist had married Herodias and Fathered Mathew, so he asked his lover Salome to get John the Baptist's head on a plate so she did a dance for Herod who was in on it and allowed her a wish. And John the Baptist was killed even though he was innocent of his incest. So Elizabeth John the Baptist's mother killed Salome for revenge due to her son's death. So the Virgin Mary slaked Alphy and she made the lord Jesus. So Mary Alphaeus Slaked Alphy and she made tripletts James the Lesser, Simon and Judas Thadeus, So Cleopus had Mary Alphaeus and made Mary.

Third Star the Sword of Judas

So Judas Thadeus found out Herod Antipas had arranged to kill John the Baptist so his sword slew Herod Antipas. So Herod Phillip Slew Judas Thadeus Brother Simon for Revenge on his brother even though Simon had not done anything wrong and was innocent, So Judas Thadeus Slew Herod Phillip to revenge Simon, so Herod Agrippa Slew James the Lesser for revenge on his brother even though James the Lesser did not do anything and was innocent so Judas Thadeus slew Herod Agrippa.

Fourth Star the Faith

So Judas Thadeus seduced Bernice to revenge his brother James the Lesser who made Mariam Arrias who married Marcus Titus Flavius Sabinus. Bernice's husband Aristobulus the father of Perpetua who married Peter, Aristobulus slew Judas Thadeus to revenge her seduction. So Peter

slew Aristobulus Bernice's husband to revenge
Judas Thadeus so Marcus Titus Flavius Sabinus
slew Peter so Andrew slew Marcus Titus Flavius
Sabinus to revenge Peter so Gaius Sillius
Calpurnius Domitius Piso slew Andrew to revenge
his father so to revenge his sons Gaius Sillius
Calpurnius Domitius Piso was slewed by Jonas.

Fifth Star the Betrayal

So Jonas had Mariamme Caecina Arria Sabinus who
had Tripplets John Mark, Miriam, and Phillip, then
Arrius Antonius Calpurnius Piso aka Flavius
Josephus was born from Jonas with his twin
Barnabus from Mariamme Caecina Arria Sabinus.
So Bonionia Prossilla Servila got Gaius Sillius
Calpurinus Piso to slew Jonas for cheating on his
wife. So Peters mother slew Bonionia Prossilla
Servila for Revenge on the death of Jonas because
Bonionia Prossilla Servila was the wife of Flavius
Josephus and Peter's mother arranged to get him
captured by Romans because she thought he was
son of Gaius Sillius Calpurnius Piso the husband of
Mariamme Caecina Arria Sabinus
So Bartholomew hid Peters Mother so Arrius
Antonius Calpurnius Piso tried to Convince the
Romans that Bartholomew was not a guilty killer
but the Romans Slew Bartholomew even though he
was innocent. So the battle began between the
Romans and James the Greater who slew the guilty
Romans including Gaius Sillius Calpurnius Piso, So
Marcus Aurelius Verus slew James the Greater so
Phillip the brother to John Mark slew Marcus
Aurelius Verus so Lucius Aurelius Verus slew Phillip
so Thomas slew Lucius Aurelius Verus so Marcus

Aurelius slew Thomas.   Judas Iscariot slew himself to become a rabbi and the Evangelist John fled to Greece, where he was sentenced to die in prison and he became an author of an apocalypse.

Works

Mathew a gospel
John Mark a gospel
John a apocalyse
Flavius Josephus a history

[106] Spell to Destroy Judah

O Great King Jeroboam strengthen King Jehu father of Azariah Father of the Mighty Helez to send Judah to Hell.

[107-110] Four Spells of the Underworld That Works Best In Hades

(107)   Make Dark Pen O God Cronos Let the Spirit of Your Divine Son Penuel Infuse Itself into this Scribe

(108)   The Spell of the Hiring

O Great Divine son of Kronos let Man hire your beautiful daughter Tzelequi Zleleponi to make the Great Divine Sampson

(109) The Spell of the Killing man

O Great Divine Son of Kronos, let your son's spirit inhabit for the Deed to be Done

(110) The Spell of the Castration

O Great Diving Son of Kronos, let your dark Scriber Great Grandson father the Sibbecai Mebunnai for the Jewels of a man

[111] Spells of Banishment

O Great King Josiah take thy wife Zebidah mother of Jehoiakim who marries Nahushta. Great Lady of Peace Nahushta rise to the Israelite Gad, whose Grandson Hagri makes the Bani.

[112] Spells of the Greatest Love

O Great High Priest Aaron guide us to your Law Giving Brother Moses whose Father Amram' Brother Great Descendent Assir has a Great Descendent Assir whose son Kore thru Obed Edom is father of Sacar the Father of the Greatest Love Ahiam.

[113-114] Spells of the Divine Good Pens

(113) O Great God Israel, let your Son Guide us in Spirit to Father Shasak who makes the Great Divine Good Scriber.

(114) O Great Archangel Gabriel, let your Son Guide us in Spirit to Father Javan who makes the Great Divine Angel Scriber.

[115] Spells of Obtaining Wealth

O Great Divine Son of Kronos, let your Grandson's Bow light up the Sky and lead us to the Pot of Gold.

[116-121] The Spells of Navigation to Heaven

(116) O Great God, Take me to David, and then his Great Grandson Shamariah, Father of Elkanah.

(117) O Great God, Take me to Blessed Feardach, and arise to Heber Hyperion who marries Tamar Granddaughter of Josiah

(118) O Great God, Take me to Neri, and arise thru his wife to her Grandfather Josiah

(119) O Great God, Take me to Levi, and then thru his son Gershom to Asaph, then to Mica of the line of Elkanah.

(120) O Great God, Take me to David, and then thru his son Sheptial thru Daniel to Achbor

(121) O Great God, Take me to Zerubabel, and then to his daughter Shelomith, who marries Elnathan.

[122-123] The Spells of Navigation to Hades

(122) O Great Divine Son of Kronos, Take me to Lydia, and let the arc rise thru your daughter Megara to your great kingdom

(123) O Great Divine Son of Kronos, Take me to King David, and let me arise past his beautiful wife Bathsheba into your kingdom.

[124] Spell to not be able to orgasism

Adam son of God remember your father Enkido who marries Ninmah daughter of Anu God of the Sky son of Anskar son of Tiamet Goddess of sweet water who marries Apophis father of a daughter who marries Yam father of Iluyanka.

[125-126] Spell to Make Erection

(125) Xerxes and Esther Jair know your Mordecau Jair know Ira

(126) Judah father your Son who fathers Mahari and your Son fathers Baanah who fathers Helad, and then your Son fathers Ikkesh Father of Ira

[127] Spell to guard General Dodo

O Good King David let your might strengthen Elhanan mighty gurad to his father Dodo

[128] Spell to guard General Issac-har

O Good King David let your might strengthen Igal son of Joseph son of the son of Issac-har

[129] Spell to make wife for a Lord

O Good King David let your might strengthen Zelek.   Let Abala know Nahash and father Shobi Father of Zelek Father of a wife for a Lord

[130] Spell to Hire for a Work

From Cainen Abimilkas Father Ahiram Father of Abibaal

[131] Spell to guard the Goddess Bathsheba

O Good King David let your might strengthen Eliam.   Bathsheba when in trouble bring your Grandfather Athisophel's mighty son to your aid, Eliam.

[132] Spell to Calm a Women

ArchAngel Michael Find your arc from Gad, and let Mary know Hyper duluth and let the women rise to Heber Hyperion whose wife Grandfather is Good King Josiah. Let the Good King Josiah know his wife Zebudah and let her take the women to her father Pediah son of Jehiochin son of Queen Nahushta. Let the Good Queen take the women up her line to Elkanah thru his wife to the shelter of Gilead son of Michael.

[133] Spell of Fear

O Great God Kronos let your GreatGrandson's wives Spirit Infest this man

[134-141] Spells of Navigation to a Higher Heaven

(134) O Great Divine Son of Zeus Apollo let our son Ion lead to his grandson Pherecydes whose Great Great Grandfather marries Tiamet Goddess of Ocean of Sweet Water

(135) O Great King Nechebenezzar Live Forever and let your ancesters the Great Sargon's line rise to The Great Gilgamesh son of The Goddess Astarte of Love, War, And Fertility

(136) O Great Warrior Uriah let your line rise to an unknown son of Nimrod, son of the Goddess Astarte Goddess of Love, War, And Fertility.

(137) O Greatest Warrior Hercules find your Great Grandfather Perseus who marries Andromeda whose Great Grandfather the Great God Poseiden's wife Libya Grandfather Telegonus is many generations the son of Kenkene The son of Gilgamesh the Son of Astarte.

(138) O God let your lines to your son Adam and his wife Eve lead to your wife Ninmah the daughter of The God of the Sky Anu

(139) O Father of the Hebrews let your wife Sarai's line lead to Boethus son of Nimrod who married the Goddess Astarte Goddess of Love, War And Fertility.

(140) O Father of the Hebrews let your concubine Hagar line rise to her mother Zeptah daughter of Onitah son of Nimrod who married the Goddess Astarte Goddess of Love, War And Fertility.

(141) O Beautiful Daughter Lulu-wa let your mother Lilith the Beautiful Queen marry Enkidu son of Eresh-Kigal daughter of Nanna Sin son of Enlil God of the Wind the Jealous God son of Anu God of the Sky

[142] Spell To Enter the Heaven Olympus

Our Lady of the Cross lead us to the Worthy Warrior Emperor Alexander the Great and let his mother Olympias line rise to Neoptolemus son of Achilles and Iphigenia daugher of Helen

www.ingramcontent.com/pod-product-compliance
Lightning Source LLC
Chambersburg PA
CBHW081144180526
45170CB00006B/1926